ELL Activity

Grade K–2

GO MATH!

PROVIDES Correlations to Chapter Content

INCLUDES Copying Masters

Printed in the U.S.A.

ISBN 978-0-544-40103-7

10 0982 23 22 21 20 19 18

4500727008 ^ B C D E F G

Contents

How to Use the ELL Activity Guide

The ELL Activity Guide is designed to supplement English language acquisition for English Learners. The activities are designed to help children acquire math vocabulary and the language and writing skills necessary to communicate and understand math concepts. The activities will also help children progress through the language proficiency levels to gain fluency.

The ELL Activity Guide also provides support for assessing your children as they progress through the language proficiency levels and for providing effective teaching for English Learners.

Strategies for Effective Language Teaching

Create a Supportive Environment English Learners come from different linguistic, cultural, and social backgrounds, and bring a variety of educational needs to the classroom. Teachers need to be sensitive to those needs and establish an environment where all children feel safe and free to take risks.

Create Comprehensible Input Teachers can provide comprehensible input through the use of gestures, appropriate speech rate, dramatization, visuals, realia, and hands-on activities using manipulatives, charts, repetition, and rephrasing.

Encourage Student Interaction Because of the social nature of language learning, children need continual opportunities for interaction, both child-to-child and teacher-to-child. Provide English Learners with frequent opportunities to interact with native English speakers.

Build Background Working with examples and problems drawn from the various content areas requires children to constantly build on their knowledge base. Teachers need to ask, "What prerequisite knowledge do my children need in order to comprehend and learn this new material?" Video clips, pictures, magazines, trade books, and printed materials in children's primary language can all be used to provide the background knowledge needed for success.

Tap Prior Knowledge English Learners bring unique experiences and backgrounds to the classroom. To assist them in building meaning from these experiences and backgrounds, teachers need to help them connect what they know to new concepts being taught.

Teach Study and Reference Skills To be successful in learning content-area material, children need organizational skills. Using graphic organizers to gather and organize information helps children see comparisons, connections, and relationships between concepts, ideas, and words.

Provide for Primary Language Support English Learners who do not receive formal content-area instruction in their primary language need support. Teachers can use peer and cross-age tutors as well as parents and community volunteers. If a paraprofessional is available to provide primary language support, have them preview the upcoming lesson in the children's primary language. After the lesson is taught by the teacher in English, have the paraprofessional review the lesson to identify any misunderstandings that could be related to language barriers.

Instruction, not content, needs to be modified to meet the needs of English Learners. Content must remain rigorous.

Effective Instructional Strategies

The activities in the ELL Activity Guide are based around instructional strategies that will help teachers assess children's mathematics knowledge and help children communicate about math on their current language proficiency level.

These instructional strategies are as follows.

Illustrate Understanding Children illustrate or sketch to communicate their understanding. Children can share their understanding through pictures and acquire new vocabulary by matching words to picture definitions.

Scaffold Language Through guided modeling and questions, children communicate understanding of a thing, situation, or process to demonstrate verbally or in writing what they already know or what they have learned.

Identify Relationships Children identify relationships between concepts and words to connect new information to prior knowledge.

Understand Context Children identify and experience vocabulary in context to clarify the meanings of words, idioms, and colloquial expressions. They can also explore synonyms and homophones to help distinguish math words from everyday words that have a similar meaning or sound the same.

Develop Meanings Children define words by using them in context with definitions or by matching words or visuals to their definitions.

Rephrase After hearing or reading a passage, children rewrite or verbally rephrase the passage in their own words. Children demonstrate their summary skills and ability to find important information.

Model Language Spoken and written language is demonstrated for children and repeated by children to practice pronunciation, sentence structure, and spelling.

Language Proficiency Levels

Researchers have identified three stages of development through which learners of a language progress.

Emerging Children at the Emerging level must be provided substantial linguistic support. They may understand some English, but they won't speak or write very much. Teachers can check understanding by asking children to point to things, use other gestures, or draw. As their skills develop, children begin to produce one- or two-word answers voluntarily. They are still building vocabulary, but to a question such as "Who is older, Carlos or Cathy?" they can answer, "Carlos."

Help children progress through the Emerging level by giving simple, clear directions and using gestures and facial expressions to convey meaning. Be sure to model correct English sentence structure and pronunciation and provide many hands-on experiences.

Expanding At this level, children begin to use short sentences. If the teacher asks, "Which number is greater, 96 or 54?" children may answer, "96 is greater." Children still use imprecise language and many nonstandard forms, but they say more, and they are moving toward standard English.

Children at the Expanding level can follow more complex directions. They need opportunities to use academic math vocabulary and concepts in interactive small- and whole-group discussions.

Bridging Children at the Bridging level are able to communicate in ways that are appropriate to different tasks, purposes, and audiences. They are transitioning to full engagement in grade-level academic tasks and activities without the need for specialized instruction.

At the Bridging level, children will be able to read and write with some fluency. They should be encouraged to justify and explain their work using academic language.

Keeping the levels of English proficiency in mind can help teachers tailor instructional practices so that children have the best possible environment for learning English.

Assessing Language Development

No single tool can accurately measure a child's level of language proficiency and development. Frequent, informal observation of child behavior and work is necessary to form an accurate picture. The following tips and tools can help you observe and evaluate child performance.

Informal Assessment Tips

- Allow **sufficient time** for children to respond.
- If a child's response is inappropriate or inadequate, try **rephrasing** your question.
- Observe children in **situations** in which they are **comfortable.** For example, some children are more comfortable in small groups than in whole-class situations.
- Observe children at **different times**, on **different tasks**.

Informal Assessment Tools

Keeping **Math Journals** helps children practice their communication skills and provides teachers with a valuable resource to assess those skills. Encourage children to record questions, understandings, examples, and illustrations.

Student Assessment Checklist

The checklist on pages viii–x will help you **evaluate** children at different levels of language development.

Periodically use the checklist to evaluate a child's current language proficiency level to match children to the activities they should complete.

Student Assessment Checklist

Child's Name ______________________________ **Date** ______________

To evaluate a child's progress, observe behaviors at different times. Use a plus (+) or minus (−) symbol to indicate that the child has or has not exhibited the behavior.

		+ or −	Comments
	Emerging		
Listening	Comprehends frequently occurring words and phrases		
	Follows simple directions		
	Comprehends some whole-class instruction		
	Begins to follow group discussions		
Speaking	Responds nonverbally by gesturing and imitating		
	Uses everyday, basic vocabulary		
	Makes short, appropriate oral responses to questions		
	May exhibit frequent errors in pronunciation and grammar		
Reading	Reads brief texts with simple sentences, supported by graphics or pictures		
Writing	Illustrates to convey meaning and ideas		
	Writes familiar words or phrases		
	Writes collaboratively		

Student Assessment Checklist

Child's Name ______________________ **Date** __________

To evaluate a child's progress, observe behaviors at different times. Use a plus (+) or minus (−) symbol to indicate that the child has or has not exhibited the behavior.

	Expanding	+ or −	Comments
Listening	Understands most of what is said, in contextualized settings		
	Follows a series of directions		
Speaking	Responds using short sentences		
	Initiates simple conversations		
	May exhibit fairly frequent errors in pronunciation and grammar		
Reading	Reads texts with simple sentences independently		
Writing	Writes expanded vocabulary to provide extended responses		
	Writes short sentences		

Student Assessment Checklist

Child's Name ______________________________ **Date** ______________

To evaluate a child's progress, observe behaviors at different times. Use a plus (+) or minus (–) symbol to indicate that the child has or has not exhibited the behavior.

	Bridging	+ or –	Comments
Listening	Comprehends many abstract topics		
	Begins to recognize language subtleties		
Speaking	Expresses increasingly complex ideas using more elaborated discourse		
	Formulates and asks questions		
	May exhibit some errors in that do not usually impede meaning		
Reading	Reads increasingly complex text at grade level		
	Reads technical text supported by pictures or graphics		
Writing	Writes to meet increasingly complex academic demands		
	Writes independently		

Grade K Chapter Correlation

Use the following correlation to determine which leveled activities best fit the math in each Grade K chapter.

	Emerging	Expanding	Bridging
Chapter 1	1, 7, 10, 18, 21, 35, 50, 51	12, 56	15, 33, 42
Chapter 2	1, 18, 50, 51	12, 56	15, 33
Chapter 3	1, 7, 8, 10, 18, 30, 50, 51	12	33, 42
Chapter 4	1, 7, 10, 18, 36, 50, 51	3, 12, 32, 56	33, 42
Chapter 5	1, 7, 8, 9, 10, 18, 19, 20, 21, 50, 51	3	42, 43
Chapter 6	1, 9, 10, 18, 19, 20, 21, 30	3, 4, 12, 14, 38	15, 28, 29, 42, 43
Chapter 7	1, 10, 18, 20, 21, 30, 35, 36, 50, 51	12	42, 43
Chapter 8	1, 10, 18, 20, 21, 30, 35, 36, 50	12, 32, 56	33, 42, 43
Chapter 9	1, 11, 18, 19, 20, 21, 23	25, 40, 41, 57	15, 29, 43
Chapter 10	1, 6, 7, 8, 18, 19, 20, 21, 23, 36, 49	24, 25, 27, 39, 41, 57	15, 28, 29, 43
Chapter 11	1, 6, 19, 21, 23, 31	13, 26, 27, 55, 58	5, 15
Chapter 12	1, 11, 18, 21, 23, 50, 51	13, 25, 57	15, 33

Grade 1 Chapter Correlation

Use the following correlation to determine which leveled activities best fit the math in each Grade 1 chapter.

	Emerging	Expanding	Bridging
Chapter 1	1, 7, 8, 9, 20, 30, 35, 36, 50, 51	3, 12, 14, 37, 56	42, 43
Chapter 2	1, 18, 30, 35, 50, 51	4, 12, 14, 38, 56	28, 33, 43
Chapter 3	1, 9, 10, 18, 20, 21, 30, 50	3, 14, 37, 41, 46, 56	42, 43
Chapter 4	1, 18, 20, 30	4, 12, 14, 56	15, 28, 29
Chapter 5	1, 9, 20, 21	3, 4, 14, 37, 38, 56	15, 28, 29, 43
Chapter 6	1, 10, 18, 19, 20, 21, 50	24, 25, 41	15, 42, 43
Chapter 7	1, 18, 19, 30	32, 41, 56	33, 42, 43
Chapter 8	1, 9, 21	3, 4, 12, 37, 38	15, 28, 29, 42, 43
Chapter 9	1, 2, 18, 19, 22, 52	13, 25, 26, 27, 41, 55	5, 43, 48
Chapter 10	1, 18, 19, 20, 21, 35	25, 41, 56	15, 33, 43, 59
Chapter 11	1, 18, 19, 20, 21, 23, 36	24, 25, 41, 57	15, 43
Chapter 12	1, 11, 18, 19, 20, 21, 23, 36	24, 25, 40, 41, 57	15, 29, 43

Grade 2 Chapter Correlation

Use the following correlation to determine which leveled activities best fit the math in each Grade 2 chapter.

	Emerging	Expanding	Bridging
Chapter 1	1, 10, 18, 19, 20, 21, 23, 35, 36, 50	24, 41	15, 29, 42, 43
Chapter 2	1, 10, 18, 19, 20, 21, 23, 30, 36, 50, 51	24, 25, 32, 41, 56	15, 33, 42, 43
Chapter 3	1, 9, 10, 18, 19, 20, 21, 23, 44, 50	3, 4, 14, 24, 37, 38, 41, 46	15, 28, 29, 43, 47
Chapter 4	1, 9, 19, 20, 21, 35, 50	3, 37, 41, 46	15, 29, 33, 42, 43
Chapter 5	1, 10, 18, 21, 36, 45, 50	3, 4, 37, 38, 46	15, 28, 33, 42
Chapter 6	1, 9, 20, 45	3, 4, 37, 38, 46	29, 42, 47
Chapter 7	2, 19, 20, 21, 22, 23, 36, 51, 52, 54	25, 41	15, 16, 43, 60
Chapter 8	1, 18, 19, 20, 23, 31, 53	3, 4, 13, 24, 25, 26, 27, 41, 55	5, 34, 43, 48
Chapter 9	1, 19, 20, 23, 53	3, 4, 24, 25, 26, 27, 41, 55	5, 34, 43, 48
Chapter 10	1, 7, 8, 10, 20, 30, 35, 36, 50	41	15, 43, 59
Chapter 11	1, 6, 11, 18, 19, 20, 21, 22, 23, 36	24, 25, 40, 41, 57	15, 29, 43

partners 15 minutes

STRATEGY Illustrate Understanding

Draw to Explain

Level: Emerging
Skills: Listening, Speaking
Objective: Practice drawing and describing pictures
Materials: paper, crayons or colored markers

Activity Have partners sit together and draw a picture to explain something such as a math problem, a story, or an event. Have them use simple shapes that will be easy to discuss.

When they have finished their drawings, have them explain their pictures by asking specific questions such as **Where is the bus?** or **What shape is the sun?** Beginning students will point or respond in short answers.

Leveling: Expanding Have partners sit back-to-back and draw to explain a story or event. Have children use simple shapes that will be easy to describe.

When the children are finished with their drawings, give each child a new sheet of paper. Have partners take turns describing their drawings so that the other child can draw it without seeing it. Have the children compare drawings, discuss the differences (if there are any), and talk about other ways they could have described their drawings.

Activity 2

partners — 10 minutes

STRATEGY
Illustrate Understanding

Writing Time

Level: Emerging

Skills: Reading, Writing

Objective: Draw hands on a clock face to represent time

Materials: pencil; Blank Clocks Blackline Master, page 72

Activity Decide whether children should practice showing the time to the hour or to the half hour. Give each partner a copy of the Blank Clocks Blackline Master. Write four times children should show on their clocks.

Have partners help each other show the time on their clocks. Children draw the clock hands to show the time and then rewrite the time beneath each clock.

Partners can challenge each other with other times to show on their clocks. Make sure they draw the minute hand longer than the hour hand.

Leveling: Expanding Write the time in different ways such as "4 o'clock" or "half-past three."

Another option is to say the time and then have children draw the hands on the clock to show the time. This will also develop Listening skills.

partners 15 minutes

STRATEGY Illustrate Understanding

Add with Pictures

Level: Expanding
Skills: Listening, Speaking, Writing
Objective: Draw to show addition
Materials: paper, crayons

Activity Have partners sit together, and each draw a picture to solve an addition story problem. Have them use simple shapes or quick pictures that will be easy to discuss.

Adam takes 3 books back to the library. Sheila takes 4 books back. How many books are being taken to the library in all?	 3 + 4 = 7

After partners have finished their drawings, have one partner tell the other what was done to find the sum. Have them ask each other questions that include **How many are there in all?** or **What is the sum?** Then, have them write the addition sentence.

Extension Children can make up their own story problems. Have one partner tell the other partner a story problem. They should both draw a picture to find the sum. Have them compare drawings and explain how to find the sum.

Repeat the activity with the other partner.

Leveling: Emerging Have partners draw one picture together. Then have them answer questions by pointing to the sum, by saying the sum, or by writing the sum.

Activity 4

partners

15 minutes

STRATEGY Illustrate Understanding

Subtract with Pictures

Level: Expanding

Skills: Listening, Speaking, Writing

Objective: Draw to show subtraction

Materials: paper, crayons

Activity Have partners sit together, and each draw a picture to solve a subtraction story problem. Have them use quick pictures that will be easy to discuss.

There are 9 birds sitting on the branch.

4 birds fly away.

How many birds are left?

$9 - 4 = 5$

After partners have finished their drawings, have one partner tell the other what was done to find the difference. Have them ask each other questions that include **How many are left?** or **What is the difference?** Then, have them write the subtraction sentence.

Extension Have one partner draw a quick picture showing a subtraction problem. His or her partner tells a story problem that matches the picture. Then the first partner makes up a different story problem to match the picture.

Repeat the activity with the other partner.

Leveling: Emerging Have partners draw one picture together. Then have them answer questions by pointing to the difference, by saying the difference, or by writing how many are left.

partners 15 minutes

STRATEGY
Illustrate Understanding

Longest to Shortest

Level: Bridging
Skills: Speaking, Writing
Objective: Draw to measure and order by length
Materials: paper, crayons

Activity Have each child choose three classroom objects. Then, have partners sit together.

Ask one partner to share his or her three objects. Have both children draw a picture of each object, ordering them from *longest* to *shortest.* Make sure partners explain how they determined the order. Encourage partners to use the words "longest" and "shortest" to describe the objects.

Repeat the activity with the other partner's objects.

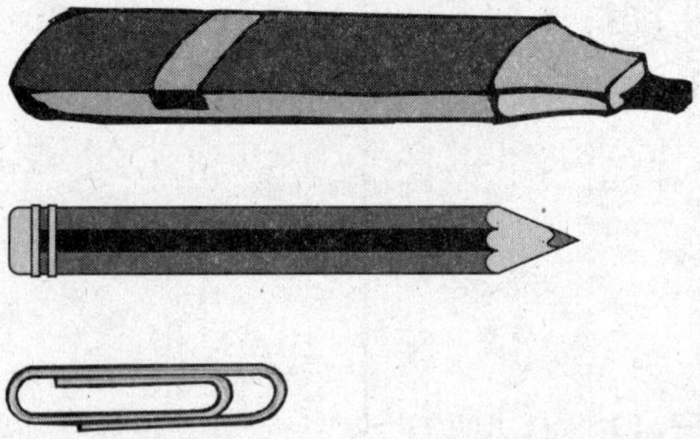

The longest object is the marker.

The shortest object is the paper clip.

After children have finished their drawings, have partners put all six objects together and order them from *shortest* to *longest.* Then have them draw pictures of the objects in this order. Have them compare and contrast ordering longest to shortest and ordering shortest to longest.

Extension Children can use color tiles to make a ruler. Line up the tiles in a row and have children trace and color the tiles. Then, have children use the rulers to measure their classroom objects. They should write the length next to each object as "about 3 tiles" or "about 6 tiles."

Leveling: Expanding Have partners only order three objects. Ask, **Which object is longest? Which object is shortest?** Children should respond in a short phrase or simple sentence.

Activity 6

small group 15 minutes

STRATEGY
Scaffold Language

Distinguish Left and Right

Level: Emerging
Skills: Listening, Speaking
Objective: Distinguish left and right
Materials: solids, paper, pencil

Activity Place a collection of solids on the table. If children have not worked with solids, use classroom objects or other manipulatives for the activity.

Name each solid. Then ask children questions: **Which solid is to the right of the cube?** pyramid **Which solid is to the right of the sphere?** cube **Which solid is to the left of the cylinder?** pyramid **Which solid is to the left of the pyramid?** cube Have children answer each question by pointing to and naming the solid.

Rearrange the solids and ask additional left and right questions.

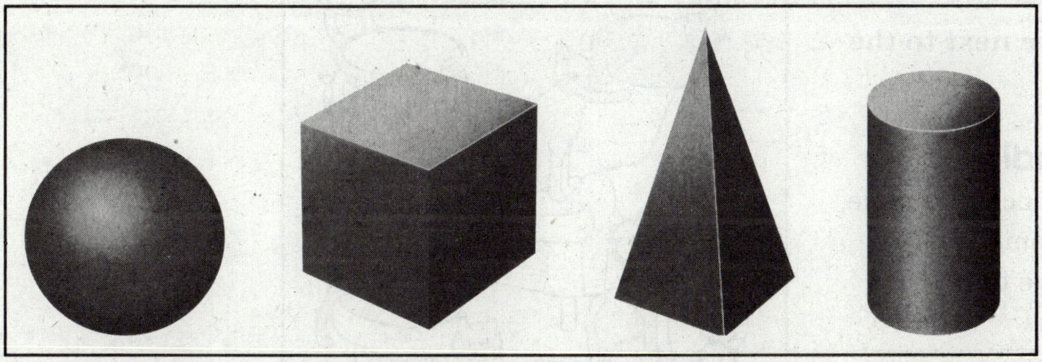

Leveling: Expanding Have children rearrange the solids. Have children ask each other left and right questions.

Then give children oral directions: **Draw a circle in the middle of your paper. Draw a square to the right of the circle. Draw a triangle to the left of the circle. Then, draw a circle to the left of the triangle.** Continue with more shapes.

partners
15 minutes

STRATEGY
Scaffold Language

Counting Cats

Level: Emerging
Skills: Listening, Speaking, Writing
Objective: Understand and use position words
Materials: Counting Cats Blackline Master, page 73; pencil; pattern blocks

Activity Give each child a copy of the Counting Cats Blackline Master. Then have children count the cats and write the number of cats.

Have children answer the following questions:

How many cats are on the chairs? 4

How many cats are below the chairs? 3

How many cats are next to the chairs? 3

Leveling: Expanding Give children 5 pattern blocks (triangle, square, hexagon, rhombus, trapezoid). Then, give the following directions.

Place the hexagon on the table.

Put the triangle above the hexagon.

Place the square next to the triangle.

Put the rhombus below the square and the trapezoid below the hexagon.

Have children use the words *on, above, below,* and *next to* to describe the blocks. Then have one child tell his or her partner where to place the pattern blocks using the words *on, above, below,* and *next to*. Then have the partner repeat the activity.

small group 15 minutes

STRATEGY Scaffold Language

Counting Monkeys

Level: Emerging

Skills: Listening, Speaking, Writing

Objective: Understand and use position words

Materials: classroom objects; Counting Monkeys Blackline Master, page 74

Activity Show children a book, holding it in front of your chest. Say, **The book is in front of me.** Hold the book behind your back. Say, **The book is behind me.**

Give each child a copy of the Counting Monkeys Blackline Master. Then have children count the monkeys and write the number of monkeys.

Have children answer the following questions:

How many monkeys do you see in all? 8

How many monkeys are in front of the trees? 4

How many monkeys are behind the trees? 4

Leveling: Expanding Have each child hold an object either in front of or behind them. Then have children tell where each object is located, using the words *in front of* and *behind.*

Have children take turns giving directions to the group members about where to put their object, using the words *in front of* and *behind.*

partners 10 minutes

STRATEGY
Scaffold Language

Spin and Add

Level: Emerging
Skills: Speaking, Writing
Objective: Add 1-digit, 2-digit, or 3-digit numbers
Materials: spinner, paper, pencil

Activity Have partners use a spinner. One child spins the spinner. Both children write down the number and say it aloud.

Then the other child spins the spinner. Both children write down the number and say it aloud. The second child adds the numbers.

Children take turns adding the numbers.

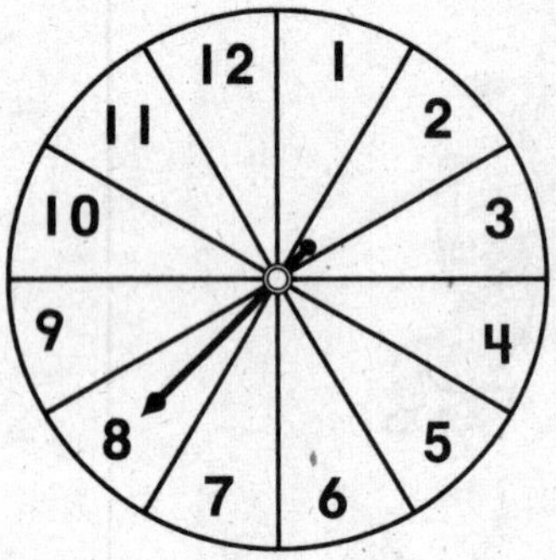

$$\begin{array}{r} 8 \\ +\ 3 \\ \hline 11 \end{array}$$

For 2-digit addition, use a spinner with the numbers 0 to 9. The first child spins two numbers to generate a 2-digit number. Then the second child spins two numbers to generate a second 2-digit number. The same can be done for 3-digit numbers, with each child spinning three times.

Leveling: Expanding Have children say the number sentence before adding the numbers. Then after the second child adds the numbers, have them say the number sentence and sum after they write it.

Activity 10

partners 10 minutes

STRATEGY
Scaffold Language

Counting

Level: Emerging
Skills: Listening, Speaking, Reading
Objective: Count objects
Materials: pattern blocks, classroom objects

Activity Have children count rows of pattern blocks or classroom objects. For pattern blocks, arrange some pattern blocks in 5 rows—the blocks in each row should all be the same shape.

Have children count each row of shapes. Ask questions about the rows of shapes. **How many squares are there?** 6 **How many trapezoids are in the last row?** 3 Continue with other questions to have children verbalize the number of shapes in each row.

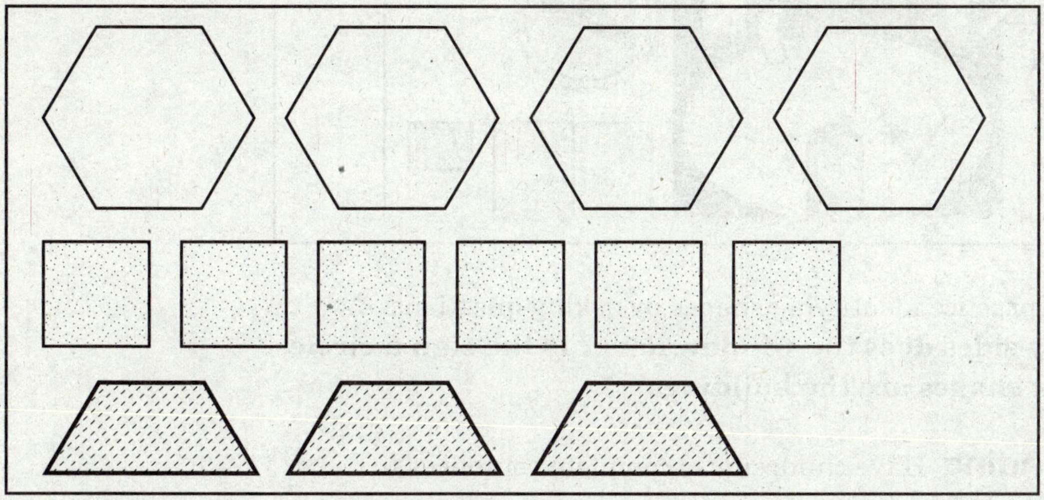

Write the words for the numbers and shapes, and read them aloud. Have children match the rows of shapes to the correct words for the number and shape.

Leveling: Expanding Allow partners to ask each other questions about the number of shapes or objects in each row. Have the partner answer in a complete sentence. Children may need a sentence frame to help them verbalize the sentences.

small group 15 minutes

STRATEGY
Scaffold Language

City of Shapes

Level: Emerging
Skills: Listening, Speaking
Objective: Identify shapes
Materials: paper, pencil, crayons

Activity Use simple shapes to draw a city like the one shown.

Then have children identify or describe several shapes in the drawing by pointing or giving short answers; for example, the sign is a circle, the buildings are square, the house has curves.

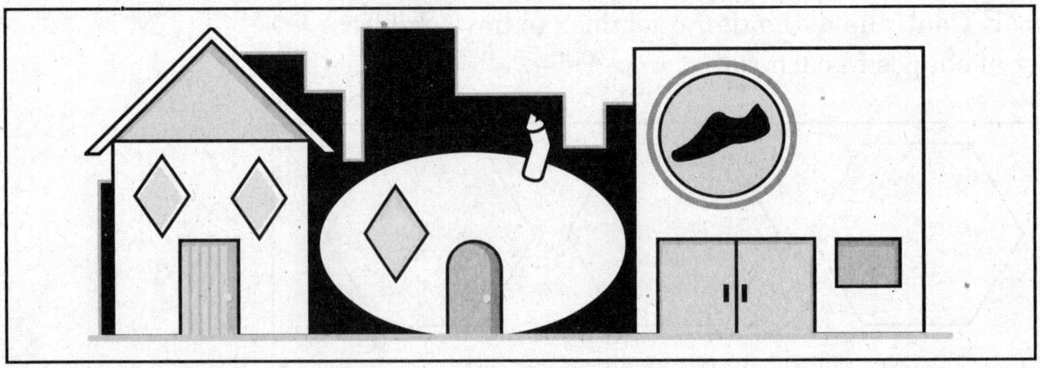

Give children more practice identifying shapes by asking questions such as, **How many sides does the window have? Is the sign a circle or a square? What shapes are the buildings?**

Leveling: Expanding Have children use complete sentences to identify where they see each shape.

Then, have children use the shapes to draw a city of their own. Have them ask each other questions about the shapes in their cities.

Activity 12

small group · 15 minutes

STRATEGY
Scaffold Language

Musical Chairs

Level: Expanding
Skills: Listening, Speaking
Objective: Count and compare numbers
Materials: one fewer chair than children in the small group, music CD, CD player

Activity Write the sentence frames. Have children complete the sentences.

Then have children play musical chairs. Place the chairs back to back. Children march around the chairs as the music plays. Tell children to sit in a chair when the music stops. The child left standing is eliminated from the game.

How many children are in the group? **There are _______ children.**	**How many chairs are there?** **There are _______ chairs.**
How many fewer chairs are there than children? **There is _______ fewer chair.**	**How many children are left?** **There are _______ children left.**

After a child is eliminated from the game, complete the sentence frames again to see how they change. Continue the game until only one child remains.

Extension Have children write a number sentence to show each word sentence. If there are 7 children and 6 chairs, children will write 7 − 6 to show that there is 1 fewer chair than there are children, and then after the music stops, they will write 7 − 1 to show how many children are left.

Leveling: Emerging Allow children to use pictures, gestures, or short answers instead of having children say the complete sentence.

Activity 13

small group 15 minutes

STRATEGY
Scaffold Language

What Size Is It?

Level: Expanding
Skills: Listening, Speaking, Reading
Objective: Compare length
Materials: classroom objects

Activity Write the question, sentence frame, and word bank. Display some classroom objects.

Select two objects, such as a book and a pencil. Align one end of the objects and then point to the book and ask, **What size is it?** Have children use the sentence frame and word bank. **It is longer.**

Question: What size is it?

Answer: It is ________.

Word Bank:	shortest	shorter	longer	longest

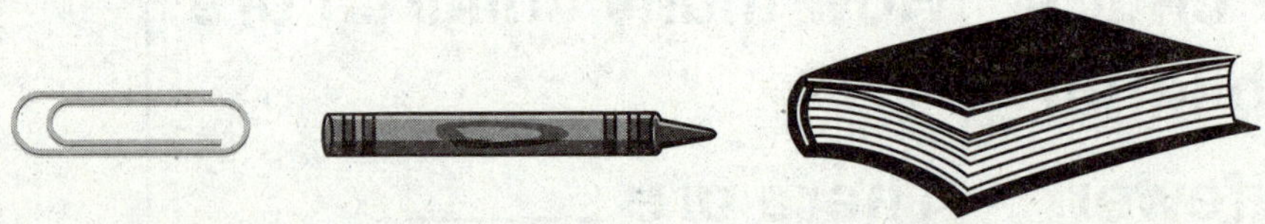

Repeat with other combinations of two and three objects.

Extension Have children choose three different classroom objects. Then have one child point to an object and ask the question. Have other children use the sentence frame and word bank to answer the question.

Leveling: Emerging Instead of using the sentence frame, allow children to use just the words in the word bank to answer the question. Children can also point to the longest, longer, shorter, or shortest object.

small group 15 minutes

STRATEGY
Scaffold Language

Fact Strategy Poems

Level: Expanding

Skills: Listening, Speaking, Reading

Objective: Practice pronunciation and vocabulary while learning fact strategies

Materials: poems Blackline Masters, pages 75–77; two-color counters

Activity When teaching fact strategies, use poems to help children understand number sentences, and understand and remember the strategies.

Give each child a copy of the poem they should recite. Read the poem with children twice, helping with any difficult words.

We Can Use Subtraction and Addition	page 75
Doubles Plus One to Get the Sum	page 76
The Sum Stays the Same	page 77

Children can take turns reading the poem in the group (with each child reciting one line as they go around in a circle), or they can read chorally as a group.

Extension Have children use two-color counters to model the strategy as they read the poem. Children can also practice writing addition and subtraction facts with numbers and symbols.

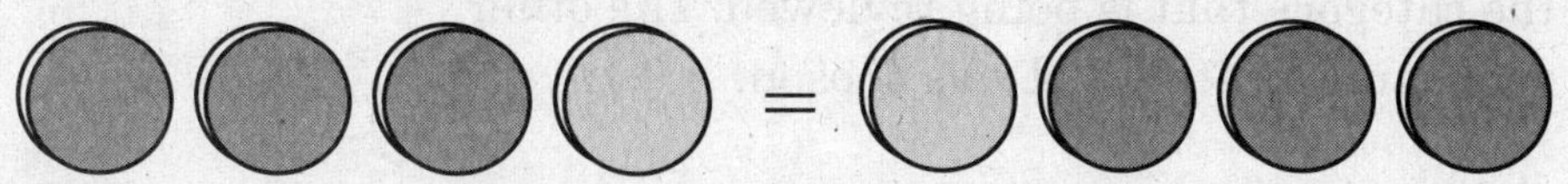

Leveling: Emerging Read the poem one line at a time and have children repeat each line. Model the concept as your read the poem.

Activity 15

small group 10 minutes

STRATEGY
Scaffold Language

Twenty Questions

Level: Bridging
Skills: Listening, Speaking
Objective: Ask questions for information
Materials: none

Activity Show children an object such as a shoe. Have them identify what it is. Then have children list words that describe the object. Write each descriptive word. Point out that they can ask questions to learn more about an object.

Review a vocabulary category such as numbers, shapes, measurement tools, school supplies, or clothing. Tell children you are thinking of a word in that category. Then guide children to ask questions to discover what word you are thinking of.

Ask questions such as:

Is your number greater than five?

Is the side of your shape straight or curved?

Can your tool hold water?

Can you write with your school supply?

What time of year would you wear the clothing?

Then have children play Twenty Questions to review vocabulary. Have a child think of a word in the category that is being reviewed. The other children ask questions to determine which word was chosen.

Extension Have children use complete sentences to answer each question. In addition, if a child guesses the correct word, have him or her explain how he or she knew what the word was.

Leveling: Expanding Have children ask questions that can be answered with a yes or no.

Activity 16

small group — 10 minutes

STRATEGY
Scaffold Language

Explain How to Add

Level: Bridging
Skills: Listening, Reading, Writing
Objective: Write an explanation for adding money
Materials: paper, pencil

Activity Have children use the sentence frame to write an explanation for adding money amounts.

Write the sentence frame explanation for children and give them the prices of five items. Children will have the most success with pictures of the items or real objects with labeled prices.

Say, **Imagine you have $15.00 to buy two presents for a friend. Choose two presents to buy for your friend. Then use the sentence frames to explain how to find the cost of the two presents.**

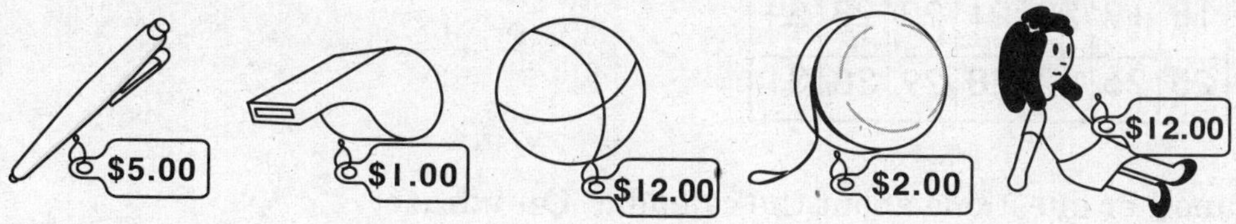

I selected _______ and _______. To find the cost of two presents, first I _______. Then, I _______. Finally, I _______. They cost _______. That is _______ than $15.00.

Write the following words that children can use in their explanations. Math words: *select, in a column, count on, sum, in all, total, greater, less*

Extension Have children select three presents that total less than $15.00.

Leveling: Expanding Allow children to use simple sentences to tell how they found the cost.

partners | 10 minutes

STRATEGY
Scaffold Language

Read a Calendar

Level: Bridging
Skills: Listening, Speaking, Reading, Writing
Objective: Understand days and months
Materials: calendar, pencil, blank calendar page

Activity Show children a calendar. Have one child explain where to find the days of the week on the calendar. Then have a child say the days of the week as he or she points to them on the calendar.

AUGUST						
S	M	T	W	T	F	S
				1	2	3
4	5	6	7	8	9	10
11	12	13	14	15	16	17
18	19	20	21	22	23	24
25	26	27	28	29	30	31

Have children ask one another questions about the calendar. **On what day of the week is the 23rd? What month comes after September?** Then, have children say the months of the year together.

Extension Give children a blank calendar page. Have children fill out the calendar together, labeling the days of the week, the month, and the numbers.

Leveling: Expanding Have children work with the calendar as a small group. Instead of having children ask questions of one another, ask questions of the group. Make sure children use short answers or complete sentences when answering the questions.

Activity 18

partners 10 minutes

STRATEGY Identify Relationships

Which Picture Doesn't Belong?

Level: Emerging

Skills: Speaking

Objective: Make connections among pictured vocabulary words to identify relationships

Materials: 4–6 pictures that illustrate vocabulary from the chapter

Activity Provide children with 4–6 pictures that illustrate vocabulary. All of the pictures should be related in some way except for one. Children choose the picture that doesn't belong.

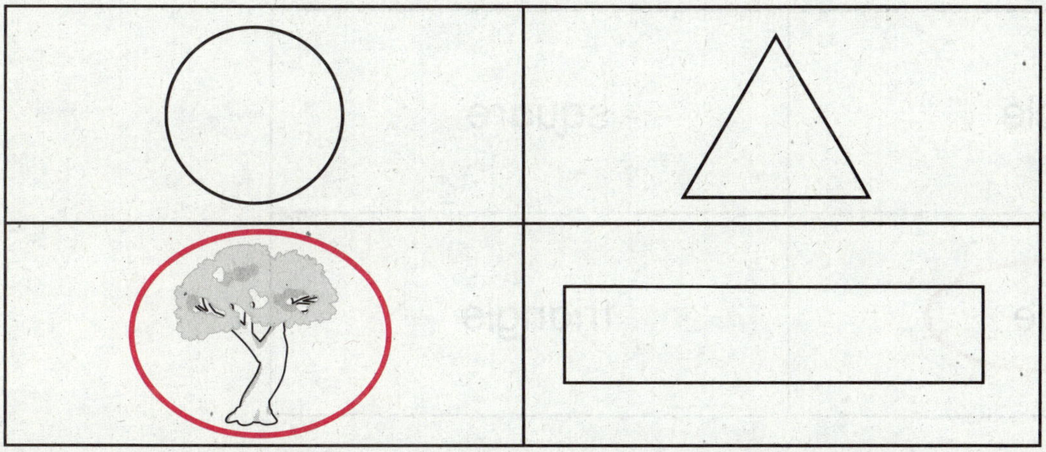

Encourage children to discuss how the other pictures are related to reinforce their decision. If they don't know the name for one of the pictures, name it for them and have them repeat the word.

Leveling: Expanding Select pictures that are more closely related, for example a nickel, a dime, a penny, and a dollar, or three clocks that show time to the hour and one that shows time to the half hour.

Have children complete the following sentence frames to help them discuss their choice.

__The tree__ does not belong.

The other pictures are __plane shapes__.

Activity 19

partners | 15 minutes

STRATEGY
Identify Relationships

Which Word Doesn't Belong?

Level: Emerging

Skills: Speaking, Reading

Objective: Practice vocabulary, making connections between new words and prior knowledge

Materials: list of 4–6 words, use the vocabulary charts, pages 61–71, or vocabulary from the chapter

Activity Provide children with 4–6 vocabulary words. All of the words should be related in some way except for one. Children choose the word that doesn't belong.

circle	square
blue	triangle

Encourage children to discuss the meaning of each word to reinforce their decision. If they don't know the meaning of a word, provide them with a definition and examples. Have partners try to explain to each other before helping them.

Leveling: Expanding Select words that are more closely related, for example rhombus, square, triangle, and rectangle.

Have children complete the following sentence frames to help them discuss their choice.

Blue does not belong.

Blue is a word for color.

The other words are plane shapes.

partners 15 minutes

STRATEGY Identify Relationships

Semantic Map

Level: Emerging

Skills: Speaking, Reading, Writing

Objective: Identify related vocabulary words by filling in a semantic map

Materials: Semantic Map Blackline Master, page 78; list of 6–10 words, use the vocabulary charts, pages 61–71, or vocabulary from the chapter

Activity Give partners a copy of the Semantic Map Blackline Master. Provide children with a topic and 6–10 vocabulary words. Have partners take turns selecting words that fit in the semantic map.

If there are fewer words, you can cover or color the extra spokes. (If there are more words, you can draw additional spokes.)

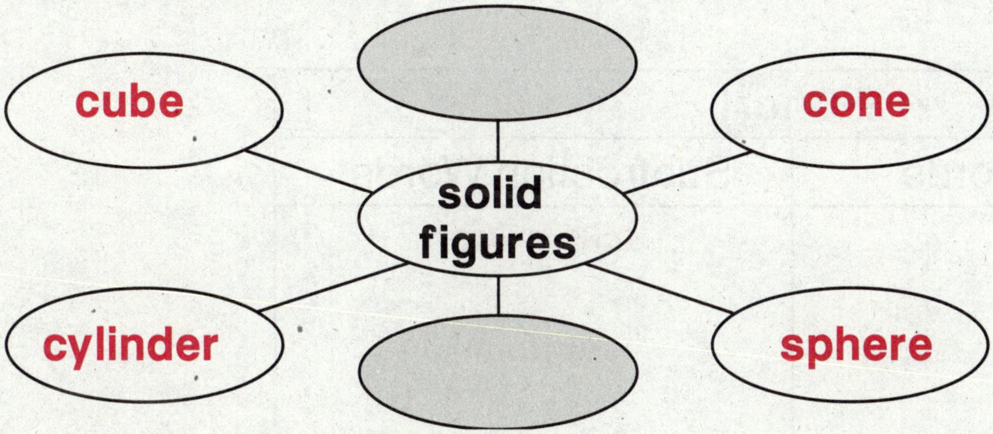

Children can copy the written words or place cards on the semantic map to show which words are related to the topic.

Leveling: Expanding Have partners think of vocabulary words that relate to the given topic. If children have difficulty thinking of some words, provide words they are studying. Allow children to explore how these new words are like the ones they thought of on their own.

Activity 2I

small group 15 minutes

STRATEGY
Identify Relationships

T-Chart

Level: Emerging

Skills: Reading, Writing

Objective: Practice vocabulary, making connections between new words and prior knowledge

Materials: list of 5–8 words, use the vocabulary charts, pages 61–71, or vocabulary from the chapter

Activity Draw a T-chart for a small group of children. Make sure they can all see the T-chart.

Provide children with 5–8 vocabulary words and two sorting categories. Some of the words should be related to one sorting category and the other words should be related to the other sorting category. Children then sort the words into the correct groups.

For example,

T-Chart	
Addition Words	Subtraction Words
add	take away
total	subtract
sum	

Encourage children to discuss the meaning of each word to reinforce their decision. If they don't know the meaning of a word, provide them with a definition and examples. Have children try to explain to each other before helping them.

Leveling: Expanding Select words that are more closely related; for example, rhombus, square, triangle, and rectangle. Provide children with 8–12 terms.

Activity 22

partners — 10 minutes

STRATEGY
Identify Relationships

Read and Write Time

Level: Emerging

Skills: Speaking, Reading, Writing

Objective: Write the time a clock shows

Materials: pencil; Time to the Hour Blackline Master, page 79, or Time to the Half Hour Blackline Master, page 80

Activity Decide whether children should practice reading the time to the hour or to the half hour. Give each partner a copy of either the Time to the Hour Blackline Master or the Time to the Half Hour Blackline Master.

Have partners help each other decide what time each clock shows. Children should write the time beneath each clock.

Have partners discuss how they knew the time. Ask questions such as, **Where does the hour hand point? Where does the minute hand point?** Help them verbalize that for time to the hour the minute hand always points to 12, and for time to the half hour the minute hand always points to 6.

Leveling: Expanding Have children say what time the clock shows. One partner should ask, **What time is it?** The other partner should respond, **It is four o'clock.**

For time to the half hour, challenge children to say the time in different ways such as "three thirty" and "half-past three."

Activity 23

partners

10 minutes

STRATEGY
Identify Relationships

Venn Diagram

Level: Emerging

Skills: Speaking, Reading, Writing

Objective: Identify words that are related to the same category

Materials: list of 6–10 words, use the vocabulary charts, pages 61–71, or vocabulary from the chapter

Activity Draw two large circles for a small group of children to make a Venn diagram.

Provide children with 6–10 vocabulary words and two sorting categories. Some of the words should be related to one sorting category and the other words should be related to the other sorting category. Children then sort the words into the correct groups.

For example,

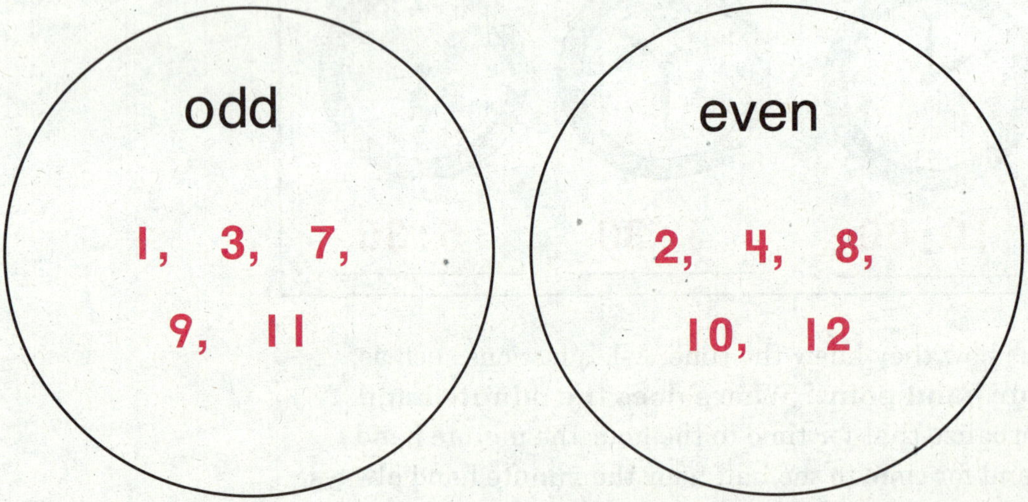

If children don't know the meaning of a word, provide them with a definition and examples. Have partners try to explain to each other before helping them.

Leveling: Expanding Have one partner sort the words by any category he or she wishes. Then have the other partner determine which categories the first partner used.

Provide children with 10–14 terms.

STRATEGY
Identify Relationships

Overlapping Venn Diagram

Level: Expanding

Skills: Speaking, Reading, Writing

Objective: Identify words that are related to one category or to two categories

Materials: list of 5–8 words, use the vocabulary charts, pages 61–71, or vocabulary from the chapter

Activity Draw two large overlapping circles for a small group of children to make an overlapping Venn Diagram.

Provide children with 5–8 vocabulary words and two sorting categories. Some of the words should be related to only one sorting category, some words should be related to only the other sorting category, and some words should be related to both sorting categories. Children then sort the words into the correct groups.

For example,

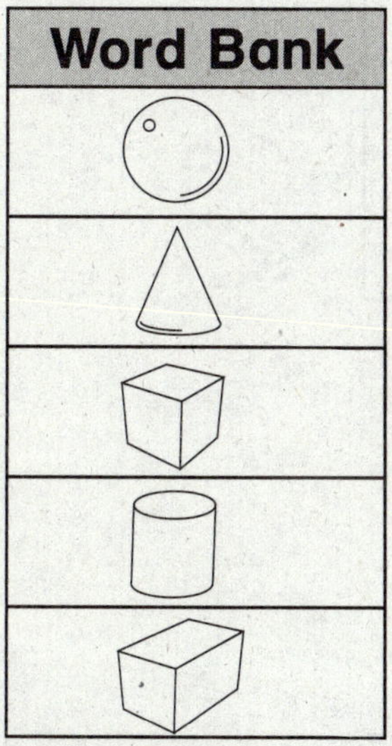

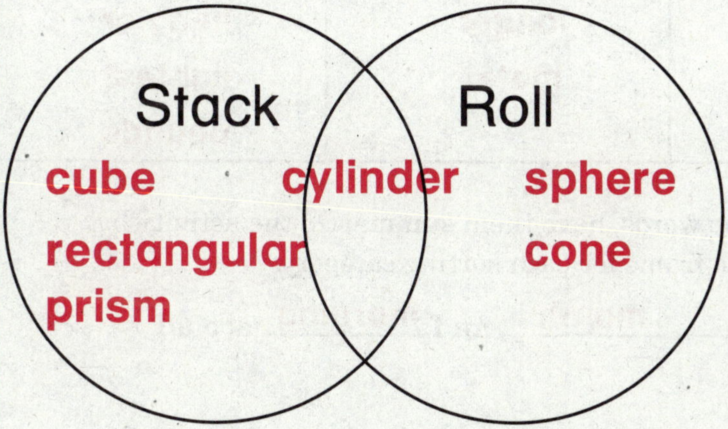

Extension Have children discuss the meaning of each word or show that the word fits a category.

Leveling: Emerging Have children sort terms that can be represented physically or by drawing. Ask questions to help children identify words that fit both categories.

partners 10 minutes

STRATEGY Identify Relationships

Let's Sort

Level: Expanding
Skills: Speaking, Reading, Writing
Objective: Sort words into three categories
Materials: list of 6–10 words, use the vocabulary charts, pages 61–71, or vocabulary from the chapter

Activity Draw a chart with three columns or three rows.

Provide children with 6–10 vocabulary words and three sorting categories. Each word should be related to only one sorting category. Children then sort the words into the correct groups.

For example,

Let's Sort		
Time	**Length**	**Weight**
day month morning	inch longer meter	grams heavier lightest pounds

After children sort the words, have them summarize the activity by completing a sentence frame for each sorting category.

The words __day__, __month__, and __morning__ are all __time__ words.

Extension Challenge children by giving them a list of words that only fit two of the sorting categories. Have them identify why none of the words fit into the last sorting category.

Leveling: Emerging Have children sort terms that can be represented physically or by drawing.

Activity 26

partners 10 minutes

STRATEGY Identify Relationships

Order by Length

Level: Expanding
Skills: Reading, Writing
Objective: Order items by length
Materials: 4–6 objects to order by length

Activity Give children a list of 4–6 objects for them to order by length.

Then draw a chart to help children order the items. The chart should have a place for the name of each item, visually showing that the items are ordered from shortest to longest.

Have children order the items and write the name of each item in the correct place on the chart.

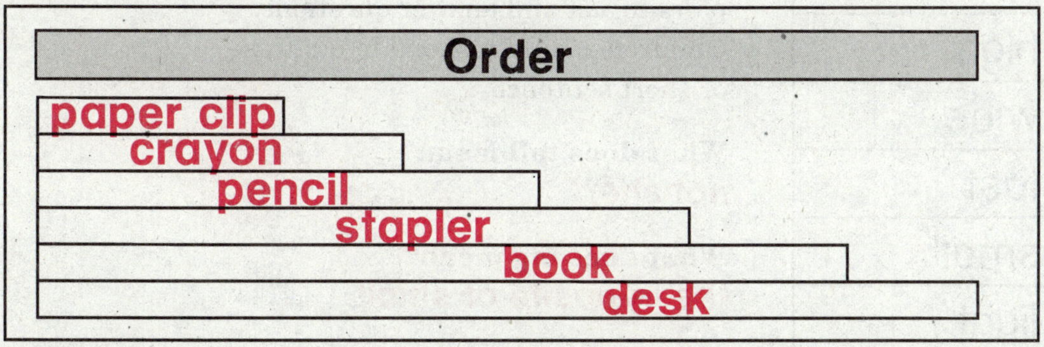

Extension After children order the objects, have them read the order by completing sentence frames to describe the relationships. Have children identify the shortest and longest items.

A ___crayon___ is longer than a ___paper clip___.
A ___desk___ is longer than a ___book___.

The ___paper clip___ is shortest.
The ___desk___ is longest.

Leveling: Emerging Have children order the physical objects and then complete the chart by writing the names of the objects.

partners 10 minutes

STRATEGY Identify Relationships

Opposites

Level: Expanding

Skills: Listening, Speaking

Objective: Make connections between words that are opposites

Materials: paper, crayons or colored markers

Activity Provide children with a list of words and their opposites. Have children match pairs of words to make connections between words that are opposites. Some examples of opposites are shown.

slow	fast
tall	short
low	high
narrow	wide
first	last
big	small
heavy	light
cold	hot

After children match the word pairs, ask children to define each word by its opposite. Have them work in pairs to ask and answer questions. Children should respond in a phrase or short sentence.

What does tall mean?
not short

What does big mean?
the opposite of small

Use other word pairs as necessary to help children learn words to describe math concepts.

Leveling: Emerging Use fewer pairs of words. Have each partner choose one word in a pair, and then have the partners act out or draw the meanings of the words to demonstrate the opposites.

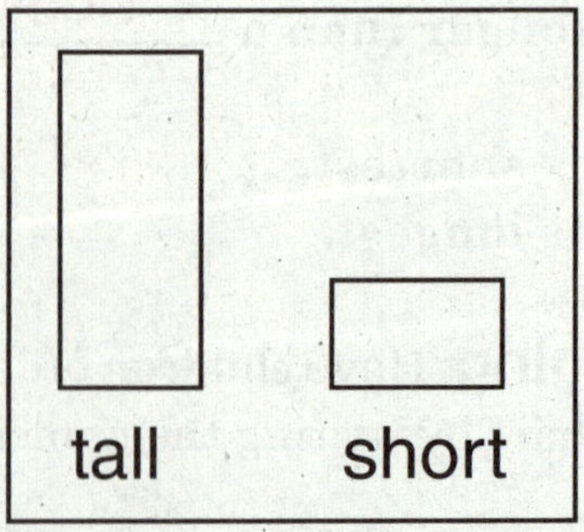

partners 10 minutes

STRATEGY
Identify Relationships

Match Related Words

Level: Bridging

Skills: Listening, Speaking, Writing

Objective: Match pairs of related words to make connections between words

Materials: list of 6–12 words, use the vocabulary charts on pages 61–71, or vocabulary from the chapter

Activity Give children 6–12 math words that can be matched in related pairs. (Some examples are provided; tailor the activity to use words children are currently studying or have difficulty pronouncing, reading, or defining.)

Have children match the word pairs that are related. Encourage children to read each word aloud.

plus	add	above	below
minus	subtract	inside	outside
sum	total	beside	next to

Have children discuss how the words are related. Suggest that they say and write each pair of words in a sentence to show how they are related. **I use a plus sign to add. The sum is also called the total. Above is the opposite of below.**

Extension Give children words that can be matched in groups of three or four. Have children identify how all the words are related. For example, total, add, and join all signify addition; morning, afternoon, and evening are all times of day.

Leveling: Expanding Provide a sentence frame so children can describe how the words are related, or dictate a sentence for children to write.

partners 10 minutes

STRATEGY
Identify Relationships

Match Related Ideas

Level: Bridging

Skills: Listening, Speaking, Writing

Objective: Match pairs of related ideas to make connections between words

Materials: 6–12 pictures of math words or ideas

Activity Give children 6–12 pictures that represent math words or math ideas. The pictures should be able to be matched in pairs of related ideas. (Some examples are provided; tailor the activity to use words children are currently studying or have difficulty pronouncing or explaining.)

Have children match the pictures that are related. Encourage children to name what each picture represents.

Part

Whole

day	night
above	below
inside	outside
full	empty
add	subtract
flat	curved

Have children discuss how the ideas are related. Suggest that they say and write a sentence to show how the ideas are related. **The apple slice is part of the whole apple.**

Extension Give children words that can be matched in groups of three or four. Have children identify how all the words are related.

Leveling: Expanding Provide a sentence frame so children can describe how the ideas are related, or dictate a sentence for children to write.

partners 15 minutes

STRATEGY
Understand Context

More and Less

Level: Emerging

Skills: Listening, Speaking

Objective: Practice identifying more and less

Materials: two-color counters

Activity Place a number of two-color counters on a workmat. Then make a second set that shows one more counter. Say, **There is <u>one more</u> counter.** Have children repeat the phrase "one more."

Change the second model so that it shows one less counter than the first set. Say, **There is <u>one less</u> counter.** Have children repeat the phrase "one less."

Make other sets of counters to show one more and one less. Prompt children to say "one more" or "one less."

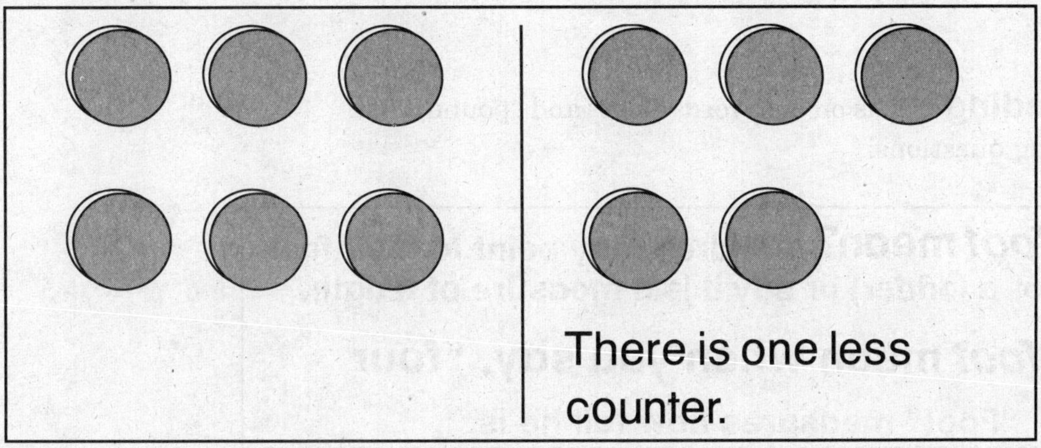

When children can confidently identify and say "one more" or "one less," make sets to show two more and two less or three more and three less.

Leveling: Expanding Ask children to make sets to show one more, one less, two more, or two less. Have partners identify what the second set shows. Write a sentence frame and have children identify more or less in a complete sentence: **There is (are) ______ ______ counter(s).**

Activity 3I

small group | 10 minutes

STRATEGY
Understand Context

About the Same

Level: Emerging

Skills: Listening, Speaking, Reading

Objective: Practice pronunciation and vocabulary by repeating and reciting a poem

Materials: *About the Same* Blackline Master, page 81

Activity Read the poem *About the Same* to children. Then read each line and have children repeat after you.

Some children may not be familiar with customary units of measure because many other countries use metric units on street signs and on labels. Relate familiar metric units of length and weight to the customary units (feet and pounds) to help children understand that feet measure length and pounds measure weight.

Provide each child with the poem *About the Same* so they can continue to work independently.

Leveling: Expanding Focus on the words "foot" and "pound." Ask children the following questions.

What does *foot* mean? Children may point to their feet or a base (*foot* of a ladder) or say it is a measure of length. **What does *foot* mean when you say, "four *foot* eight"?** "Foot" measures how tall he is.
What does *pound* mean? Children may describe "pound" as a verb, as currency, as an animal shelter, or as a measurement of weight. **What does *pound* mean when you say, "he weighs 52 *pounds*"?** "Pound" measures how much he weighs.

Talk to children about how some words have more than one meaning. Have children talk about how they can tell which meaning applies to the word by reading it in a sentence.

small group 15 minutes

STRATEGY
Understand Context

Before and After

Level: Expanding
Skills: Listening, Speaking
Objective: Practice using "before" and "after"
Materials: none

Activity Write the question and answer sentence frames on the whiteboard or on cards so all of the children can see them.

Question: Do you ____ before you ____?	**Question:** Do you ____ after you ____?
Answer: I ____ before I ____.	**Answer:** I ____ after I ____.

Start by modeling how to ask and answer each question. Say, **Do you eat breakfast before you eat lunch?** Have an aid or child who is fluent in English respond, **I eat breakfast before I eat lunch.** Children should repeat the answer so that they feel comfortable using the complete sentence.

Repeat the modeling process for the second question. Say, **Do you eat lunch after you eat breakfast?** Respond, **I eat lunch after I eat breakfast.**

Ask children if they do one activity before (or after) another. Sequence the events in the question to help children focus on the structure of the sentence and the meanings of "before" and "after."

Extension Once children can structure the sentence with no problems, list the events in the question out of sequence so children can identify which event is before or after the other.

Leveling: Emerging Write the words or show children pictures of the events as you ask the question. Children can point to the words or pictures in sequence saying "before" or "after." For example, children point to "lunch," say "after" and then point to "breakfast."

partners 15 minutes

STRATEGY
Understand Context

More and Fewer

Level: Bridging
Skills: Listening, Speaking
Objective: Practice identifying more and fewer
Materials: connecting cubes

Activity Place a number of red connecting cubes on a workmat. Then make a second set that shows a few more blue cubes. Say, **There are <u>more</u> blue cubes.** Have children repeat the sentence.

Change the model so that it shows more red cubes. Say: **There are <u>fewer</u> blue cubes.** Have children repeat the sentence.

Make a set of 9 yellow connecting cubes and a set of 3 green connecting cubes. Ask, **Are there more yellow cubes than green cubes?** Prompt children to respond in a complete sentence: **There are more yellow cubes than green cubes.** Ask, **Are there fewer green cubes than yellow cubes? There are fewer green cubes than yellow cubes.**

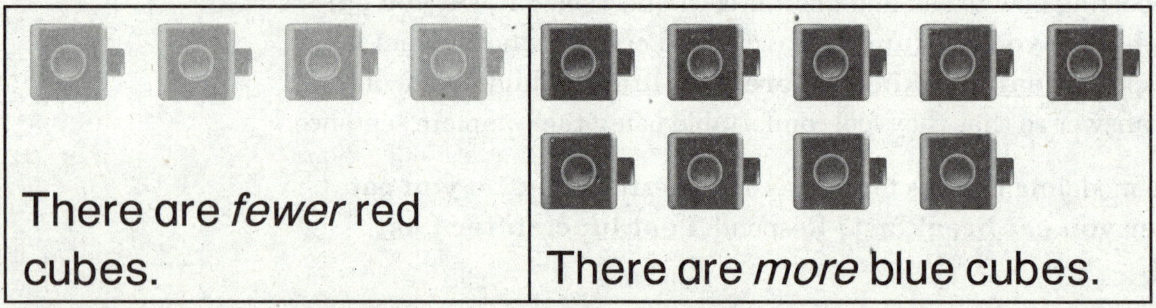

Extension Have one partner make a set of connecting cubes, then ask the other partner to make a set that shows more than their partner's set. Repeat the activity, having the second partner make a set with fewer cubes. Have children ask and identify who has more cubes: **Who has more cubes? Javier has more cubes than I have. Who has fewer cubes? Leah has fewer cubes than I have.**

Leveling: Expanding Write a sentence frame and have children identify more or fewer in a complete sentence: **There are <u>more</u> <u>green</u> cubes. There are <u>fewer</u> <u>red</u> cubes.**

small group 15 minutes

STRATEGY Understand Context

Use a Ruler

Level: Bridging
Skills: Listening, Speaking, Writing
Objective: Practice measuring length with a ruler
Materials: rulers, classroom objects that are 1 foot or longer

Activity Give children a ruler. Have them explain how to use a ruler to measure the length of a book. Encourage each child in the small group to contribute to the explanation. Write their explanation and have children copy it.

A sample explanation is provided.

> You can measure the length of a book using a ruler.
>
> First, line up the line for zero on the ruler with the end of the book. Then, count the number of inches from the zero mark to the end of the book.
>
> So, the book is about 12 inches, or one foot, long.

Have children find objects around the room that are one foot long or longer. Have children use a ruler to measure the objects. Have them say:

The ______ is ______ inches long.
The ______ is ______ foot/feet long.

Children should also record the length of each measurement.

Extension As children measure objects, have them estimate how many inches or feet long they think the object will be. Have a child record the predictions. Then, have children measure the object and record the actual measurement. Children should compare the predicted measurement to the actual measurement.

Leveling: Expanding Provide children with a list of classroom objects to measure. Then, have them work with a partner to measure the objects and record the lengths in a simple table.

partners 15 minutes

STRATEGY Develop Meanings

Matching: Everyday Words

Level: Emerging
Skills: Reading, Writing
Objective: Match words to pictures of everyday objects
Materials: pencil; Matching Blackline Masters, pages 82–85

Activity Practicing everyday vocabulary will help children understand contexts and vocabulary used in word problems.

Select a vocabulary category from one of the everyday vocabulary Matching Blackline Masters.

Matching: Animals	page 82
Matching: Clothing	page 83
Matching: Fruit	page 84
Matching: Vegetables	page 85

Give each child one or more of the everyday vocabulary Matching Blackline Masters to complete. Tell them to match each word on the left to one of the pictures on the right. Then, they write the word next to the picture.

Leveling: Expanding Have children complete the Matching Blackline Master and then write a sentence using each word. Sentences should be simple such as **I have a pet cat** or **I can buy apples.**

Ask children questions about the words such as **When do you wear a jacket?** or **Where can I buy a carrot?**

Activity 36

partners 15 minutes

STRATEGY
Develop Meanings

Matching: Math Words

Level: Emerging
Skills: Reading, Writing
Objective: Match math words to picture definitions
Materials: pencil; Matching Blackline Masters, pages 86–89

Activity Select a vocabulary category from one of the math vocabulary Matching Blackline Masters.

Matching: Numbers	page 86
Matching: Solid Shapes	page 87
Matching: Plane Shapes	page 88
Matching: Money	page 89

Give each child one or more of the math vocabulary Matching Blackline Masters to complete. Tell them to match each word on the left to one of the pictures on the right. Then, they write the word next to the picture.

Leveling: Expanding Have children complete the Matching Blackline Master and then write a sentence using each word. Sentences should be simple such as **Nine is after eight** or **A square has four sides.**

Ask children questions about the words such as **What in the classroom looks like a cube?** or **How much is a penny worth?**

Activity 37

partners 10 minutes

STRATEGY Develop Meanings

Oral Cloze: Addition

Level: Expanding
Skills: Listening, Speaking, Reading
Objective: Understand and use addition vocabulary
Materials: none

Activity Write the addition words and cloze sentences.

Have children read the sentences with the blanks. Then partners take turns completing each sentence with one of the addition words.

Addition Words: addend, addition sentence, altogether, in all, plus, total, sum

1. The answer to an addition problem is called the sum or total.
2. The symbol used in an addition sentence is called a plus sign.
3. 3 + 1 = 4 is called an addition sentence.
4. When you add, you find the sum or total.
5. Another way of saying *in all* is altogether.
6. In the addition sentence 5 + 3 = 8, the number 3 is called an addend.

Extension Display an addition sentence. Have children use the addition words to define parts of the addition sentence.

Then have children write their own addition sentence. Ask children to use the addition words to talk about their addition sentence.

Leveling: Emerging Read the sentences with blanks to children. Have them point to or say the addition word that belongs in the blank. Read the completed sentence and have children repeat the sentence.

Activity 38

partners 10 minutes

STRATEGY Develop Meanings

Oral Cloze: Subtraction

Level: Expanding
Skills: Listening, Speaking, Reading
Objective: Understand and use subtraction vocabulary
Materials: none

Activity Write the subtraction words and cloze sentences.

Have children read the sentences with the blanks. Then partners take turns completing each sentence with one of the subtraction words.

Subtraction Words: difference, minus, remain, subtract, subtraction sentence, take away

1. The answer to a subtraction problem is called the difference.
2. The symbol used in a subtraction sentence is called the minus sign.
3. 3 − 1 = 2 is called a subtraction sentence.
4. To take away is to subtract.
5. Another way of saying that no more are left is to say no more remain.

Extension Display a subtraction sentence. Have children use the subtraction words to define parts of the subtraction sentence.

Then have children write their own subtraction sentence. Ask children to use the subtraction words to talk about their subtraction sentence.

Leveling: Emerging Read the sentences with blanks to children. Have them point to or say the subtraction word that belongs in the blank. Read the completed sentence and have children repeat the sentence.

partners 10 minutes

STRATEGY
Develop Meanings

Oral Cloze: Position Words

Level: Expanding
Skills: Listening, Speaking, Reading
Objective: Understand and use position words
Materials: classroom objects

Activity Write the position words and cloze sentences.

Have children read the sentences with the blanks. Then partners take turns completing each sentence with one of the position words.

Position Words: above, below, beside, close by, far, near, next to, over

1. A position that is higher than another position is above or over.
2. A position that is lower than another position is below.
3. A position that is close by is near.
4. A position that is distant is far.
5. A position that is beside another position is next to that position.

Extension Have children use the position words to define locations of objects in the room. Have partners ask each other questions such as **What object is near the door?** or give directions such as **Hold your pencil above your head.**

Leveling: Emerging Read the sentences with blanks to children. Have them point to or say the position word that belongs in the blank. Read the completed sentence and have children repeat the sentence.

small group 10 minutes

STRATEGY
Develop Meanings

I Can Name Shapes

Level: Expanding

Skills: Listening, Speaking, Reading, Writing

Objective: Practice pronunciation and vocabulary by repeating and reciting a poem

Materials: crayons; paper; *I Can Name Shapes* Blackline Master, page 90

Activity Give each child a copy of the *I Can Name Shapes* Blackline Master. Read the poem with children twice and then have them read and say the poem independently.

Children can take turns reading the poem in the group (with each child reciting one line as they go around in a circle) or they can read chorally as a group.

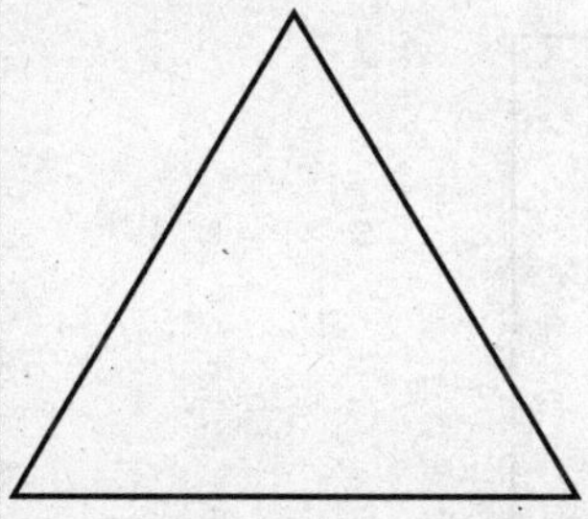

This is a triangle and I have seen it is in the color green.

Children should follow the directions in the poem to draw the shapes as they are described. They should draw a blue square, a green triangle, a red rectangle, an orange circle, and a pink rhombus.

Extension Have children draw the shapes on separate sheets of paper. Have one child read two lines from the poem and then have the other children hold up the drawing that matches those lines. After identifying the shape, have students write the lines under the drawing.

Leveling: Emerging Read the poem one line at a time and have children repeat each line. Draw the shapes as you read the poem.

small group 10 minutes

STRATEGY
Develop Meanings

Unscramble Math Words

Level: Expanding
Skills: Reading, Writing
Objective: Unscramble words to make math words
Materials: 4–5 words, use the vocabulary charts, pages 61–71, or vocabulary from the chapter

Activity Give children 4–5 vocabulary words. Scramble the letters in each word and write them on the board so that everyone in the group can see them. Draw an illustration or write an example of each word so children have a visual definition of the word they are trying to unscramble.

Children unscramble the letters to identify the vocabulary words.

Some examples are provided; tailor the activity to use words children are currently studying, or have difficulty spelling or reading.

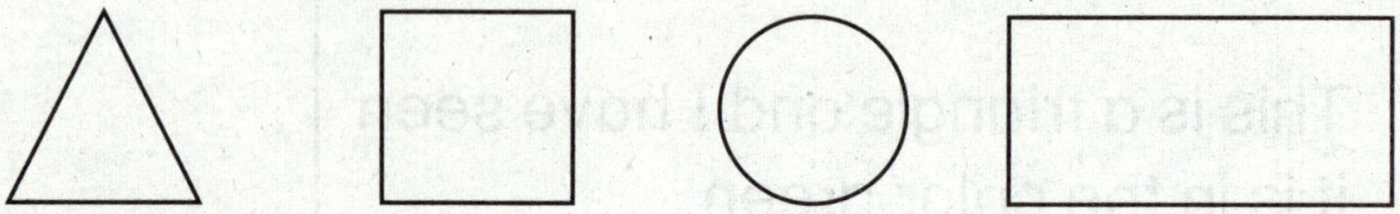

INGLEART	QAURES	CCILER	GLECTANRE
TRIANGLE	**SQUARE**	**CIRCLE**	**RECTANGLE**

Extension After unscrambling a word, have children define it in their own words.

Leveling: Emerging Provide a word bank, and allow children to look up the word and repeat the definition.

partners 15 minutes

STRATEGY
Develop Meanings

Define Numbers

Level: Bridging
Skills: Listening, Speaking, Writing
Objective: Identify numbers
Materials: base-ten blocks, paper, pencil

Activity Show children a number represented by base-ten blocks (use a one-digit, two-digit, or three-digit number to match what children are learning).

Have children write the number the base-ten blocks show. Have them use a complete sentence to identify the number.

Repeat with other numbers.

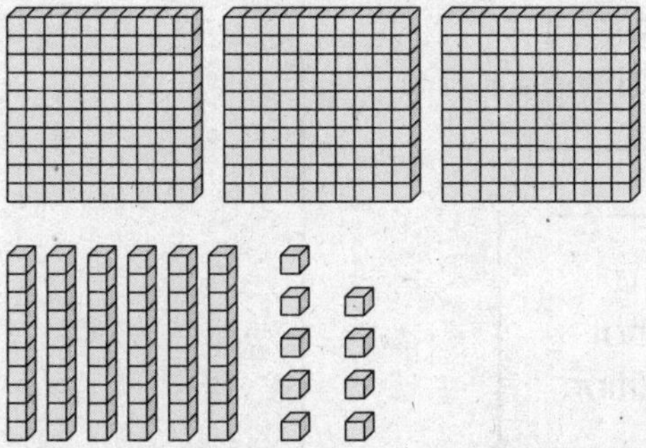

The number is three hundred sixty-nine.
369

Give partners base-ten blocks. Have them make a number using the blocks. Then have children write their partner's number. The children should come up with questions to ask their partner about his or her number such as **What digit is in the tens place?** Have them answer the questions using complete sentences.

Extension Have children make a number using base-ten blocks, but not show their partner. Then, have partners play twenty questions to find the number.

Leveling: Expanding Give children question stems to ask about numbers. Prompts can include questions such as **What digit is in the ones place?**

partners 15 minutes

STRATEGY Develop Meanings

Math Definitions

Level: Bridging
Skills: Listening, Reading, Writing
Objective: Look up, dictate, and define vocabulary words
Materials: pencil; vocabulary charts, pages 61–71, or vocabulary from the chapter

Activity Have children write vocabulary words on index cards, one word per card. One child looks up and dictates the definition while the other child writes the definition on the back of the card.

Some examples are provided; tailor the activity to use words children are currently studying, or have difficulty reading or defining.

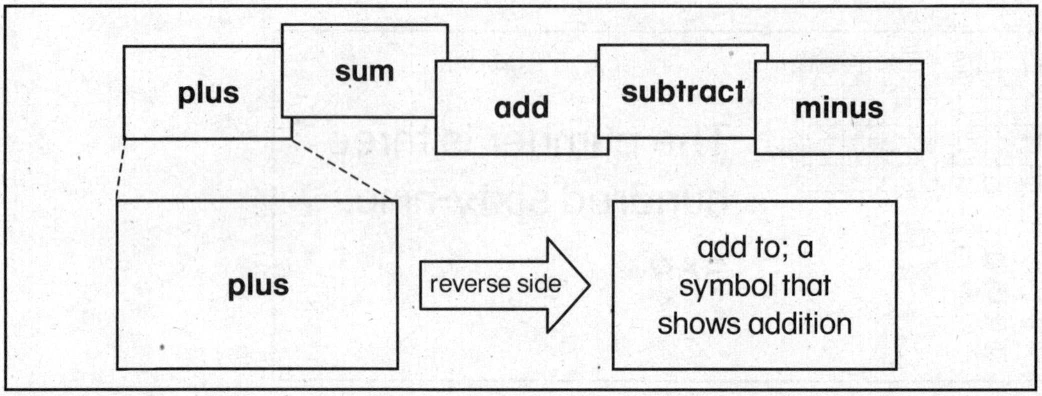

Extension Children should practice vocabulary by using these cards as flash cards.

Leveling: Expanding Read the definitions to children while they copy them on the backs of the cards. Have children draw a picture to define the word when appropriate.

partners 15 minutes

STRATEGY
Rephrase

Make a Pattern

Level: Emerging
Skills: Listening, Speaking
Objective: Rephrase the steps to make a pattern
Materials: color tiles

Activity Say the steps to make a color pattern.

Today we are going to make a color pattern using two colors.

I am going to start my pattern with a yellow tile. The next color in my pattern is blue. I will follow the blue tile with another yellow tile. Then, I use blue again.

To continue my pattern, I will place a yellow tile, and then another blue tile.

Have children rephrase the steps to a partner using their own words. Children should draw pictures or use color tiles to model the steps to make the color pattern.

Leveling: Expanding Have children change the pattern using the two colors. They can change it to an AAB pattern or an ABB pattern. You can also have them add an additional color for a three-color pattern. Have them explain the steps to make these patterns.

partners 15 minutes

STRATEGY
Rephrase

Find the Difference

Level: Emerging
Skills: Listening, Speaking, Reading
Objective: Rephrase the steps to solve a subtraction story problem
Materials: paper, crayons

Activity Display the story problem. Then state the steps to solve the problem.

Jenny has 20 invitations to her birthday party. She has given 14 of the invitations to her friends. How many invitations does she still have to hand out?

To solve this problem, I am going to draw a picture with simple shapes. I will draw 20 squares to show Jenny's invitations.

Then I will cross out 14 squares to show the 14 invitations she gave out.

Now I will count to see how many squares are left. There are 6 squares not crossed out.

My subtraction sentence is $20 - 14 = 6$. Jenny still has 6 invitations to hand out.

Have children rephrase the steps to find the difference. Have them tell their partner, in their own words, how to find how many invitations Jenny still has to hand out. Have children draw to illustrate the steps as they rephrase.

Leveling: Expanding Have children work together to make up another subtraction story problem. Then have them draw a quick picture while explaining how they will solve the problem.

small group 15 minutes

STRATEGY
Rephrase

Find the Sum

Level: Expanding
Skills: Listening, Speaking, Reading
Objective: Rephrase how to solve an addition problem
Materials: two-color counters

Activity Display the story problem. Then state the steps to solve the problem.

> **Adam has 7 balloons to blow up for his party. His mother brings him 9 more balloons. How many balloons will Adam blow up altogether?**
>
> To solve this problem, I am going to act it out using counters. I have 7 counters.
>
> Then, I will add 9 more counters to the group of 7.
>
> I will now count the total number of counters I have. I have 16 counters. $7 + 9 = 16$
>
> Altogether, Adam will blow up 16 balloons.

Have children rephrase the story problem and the steps showing how to act out the story problem and find the sum.

Write down how children rephrased the steps, and then erase a few key words. Have children read the steps and fill in the missing words.

Extension Ask children what other strategies they could have used to solve the problem. Have them explain the steps to find the sum using one of those strategies. Then, have them write a new story problem and solve it using one of the strategies.

Leveling: Emerging Have children work with a partner, taking turns to explain each step. Make sure they use the counters to act out the steps to find the sum.

small group 15 minutes

STRATEGY
Rephrase

3-Digit Operations

Level: Bridging
Skills: Listening, Speaking, Reading
Objective: Rephrase how to solve operations with 3-digit numbers
Materials: paper, pencil, base-ten blocks

Activity Display an addition or subtraction sentence. Then state the steps to break apart the numbers to add or subtract. An addition example is provided.

645 + 234 = ____

To solve this problem, I am going to break apart the 3-digit numbers.

Then I am going to add each place value together.

Finally, I will combine the place value totals to get the answer.

$$\begin{array}{rcl} 645 & \longrightarrow & 600 + 40 + 5 \\ +\ 234 & \longrightarrow & 200 + 30 + 4 \\ \hline & & 800 + 70 + 9 = 879 \end{array}$$

Have children rephrase the steps in their own words, and then tell the group how to find the answer.

Extension Ask children what other strategies they could have used to solve the problem. Then, provide children with a new addition or subtraction sentence and have them solve it using one of the suggested strategies. Have them explain the steps they used to solve the problem.

Leveling: Expanding Have children work with a partner, taking turns explaining each step. Make sure they include each step and use complete sentences. Let them use base-ten blocks or drawings to help them explain the steps.

small group 15 minutes

STRATEGY
Rephrase

Rewrite an Explanation

Level: Bridging
Skills: Speaking, Reading, Writing
Objective: Write an explanation
Materials: paper, crayons, paper clips (to make a ruler if desired)

Activity Write the explanation on the board so all of the children can see it.

Help children read the explanation for how to measure length or another explanation for the math content children are currently studying.

> Today I will tell you how to measure length.
>
> It is very easy. You will need a paper clip ruler and a pencil to measure.
>
> First, line up one end of the pencil and one end of the paper clip ruler.
>
> Then, count the paper clips from one end of the pencil to the other end of the pencil. Say, "My pencil is ______ paper clips long."
>
> Have fun measuring length!

Have children rewrite the explanation using their own words. Children may draw pictures to illustrate the steps and to aid in showing their comprehension.

Extension If children are familiar with inches, replace the paper clip ruler in the explanation with an inch ruler.

Leveling: Expanding Have children say the explanation in their own words or act it out. Write key phrases of their explanation on the board and erase some words. Have children fill in the missing words: **I need a <u>ruler</u> and something to measure. Then, I <u>line</u> up the ends.**

partners 10 minutes

STRATEGY
Model Language

Where Is It?

Level: Emerging
Skills: Listening, Speaking, Reading
Objective: Answer questions about location
Materials: drawings showing location, classroom objects

Activity Display drawings that show location or use classroom objects to have children practice how to describe location.

Point to a drawing or display an object. Ask the question and model the language children should use to answer the question.

Where is the cat**?**
It is under **the** table**.**

Then have children repeat the answer.

Children at the Emerging level may have difficulty answering the question in a complete sentence, so allow them to respond with just the location word as they gain confidence and skill. Prompt them further by asking, **Under where?** and allowing them to point to the table.

Have children identify the location of other objects in various drawings or real-word situations.

Leveling: Expanding Have partners take turns asking and answering the question. Write the sentence frames for children to work independently.

Activity 50

partners 10 minutes

STRATEGY
Model Language

How Many?

Level: Emerging
Skills: Listening, Speaking, Reading
Objective: Answer questions about how many

Materials: pattern blocks, classroom objects

Activity Arrange some pattern blocks or classroom objects in rows or groups. Each row or group should have the same shape or object.

Point to a row or group. Ask the question and model the language children should use to answer the question.

How many books are there?

There are five.

Then have children repeat the answer.

Children at the Emerging level may have difficulty answering the question in a complete sentence, so allow them to respond with just the number or with a hand gesture as they gain confidence and skill.

Change the number of pattern blocks or objects and repeat the activity.

Leveling: Expanding Have partners take turns asking and answering the question. Write the sentence frames for children to work independently.

partners 10 minutes

STRATEGY
Model Language

How Many Do You Have?

Level: Emerging
Skills: Listening, Speaking, Reading
Objective: Answer questions about how many you have
Materials: two-color counters

Activity Hold up both hands. Ask the question and model the language children should use to answer the question.

How many **fingers** do you have?
I have **ten**.

Then have children repeat the answer.

Children at the Emerging level may have difficulty answering the question in a complete sentence, so allow them to respond with just the number as they gain confidence and skill.

Have children answer questions identifying other numbers of things they have. Give them a collection of counters and then repeat the question and answer.

How many **counters** do you have?
I have **six**.

Leveling: Expanding Have partners take turns asking and answering the question. Write the sentence frames for children to work independently.

partners 10 minutes

STRATEGY
Model Language

When?

Level: Emerging
Skills: Listening, Speaking, Reading
Objective: Answer questions about when things happen
Materials: drawings or pictures showing time-specific activities

Activity Display drawings or pictures that show activities that can be done at various times of the day. Include a prompt for the time of day if the drawing is unclear or if children need to practice reading skills.

Point to a drawing. Ask the question and model the language children should use to answer the question.

When do we eat breakfast?

We eat breakfast in the morning.

Then have children repeat the answer.

Children at the Emerging level may have difficulty answering the question in a complete sentence, so allow them to respond with just the activity and the time of day as they gain confidence and skill.

Show children other drawings or name other activities and have children identify when they do those activities.

Children can also be asked about seasonal activities.

Leveling: Expanding Have partners take turns asking and answering the question. Write the sentence frames for children to work independently.

partners 10 minutes

STRATEGY
Model Language

How Far Is It?

Level: Emerging

Skills: Listening, Speaking, Reading

Objective: Answer questions about how far away something is

Materials: drawings that show distance, classroom objects

Activity Display drawings that show distance or place classroom objects to have children practice describing how far away things are.

Point to a drawing or display two objects. Make sure to give the distance or have children measure the distance. Ask the question and model the language children should use to answer the question.

How far away is the school?

It is three blocks away.

the school
3 blocks

Then have children repeat the answer.

Children at the Emerging level may have difficulty answering the question in a complete sentence, so allow them to respond with just the distance as they gain confidence and skill.

For classroom objects, have children answer the question, **How far away are the pencil and the crayon?** If they have not yet learned to measure length, children respond that the objects are **near** or **far**.

Leveling: Expanding Have partners take turns asking and answering the question. Write the sentence frames for children to work independently.

partners 10 minutes

STRATEGY
Model Language

How Much Money?

Level: Emerging
Skills: Listening, Speaking, Reading
Objective: Answer questions about money
Materials: play coins or bills

Activity Give children a collection of coins or bills. Have children arrange the coins by type.

Ask the question and model the language children should use to answer the question.

How much are the coins worth?

The coins are worth twenty-one cents.

Then have children repeat the answer.

Children at the Emerging level may have difficulty answering the question in a complete sentence, so allow them to respond with just the value of the coins as they gain confidence and skill.

Encourage children to count the value of the coins, starting with the coin of the greatest value. Model how to skip count the coins.

Leveling: Expanding Have partners take turns asking and answering the question. Write the sentence frames for children to work independently.

Activity 55

small group 15 minutes

STRATEGY Model Language

How Tall? How Old?

Level: Expanding
Skills: Listening, Speaking, Reading, Writing
Objective: Ask and answer questions about height and age
Materials: pencil, paper

Activity Help children in a small group order themselves by height. Have them sit in that order and write their names in order starting with the tallest person.

Write the sentence frames and have children ask and answer questions about height.

Peter	Who is taller than Juana?
Juana	Peter is taller than Juana.
Djibo	Who is shorter than Juana?
Susan	Djibo is shorter than Juana.

Have the other group members write the questions and answers as they are spoken.

Extension Children can also be ordered by their ages. When children are ordered by their ages they should ask and answer the questions, **Who is older than** Juana**?** and **Who is younger than** Juana**?**

Leveling: Emerging Read and model the question and answer. Children will repeat the answer.

Children at the Emerging level may have difficulty answering the question in a complete sentence, so allow them to respond with just the name of the child as they gain confidence and skill.

partners 15 minutes

STRATEGY Model Language

Compare Numbers

Level: Expanding
Skills: Listening, Speaking, Reading, Writing
Objective: Ask and answer questions to compare numbers
Materials: index cards, markers

Activity Have each partner make a set of 10–15 number cards, using numbers that children are studying.

Write the sentence frames. Have children shuffle the cards and place them in a stack. Each partner draws a card and places it on the table. Partners then compare the numbers by using the sentence frames.

Are __10__ and __10__ the same?
Yes, they are./No, they aren't.

Is __10__ greater than __5__?
__10__ is greater than __5__.

Is __10__ less than __5__?
__5__ is less than __10__.

Extension Partners can also practice addition and subtraction by adding and subtracting the two numbers they draw and by completing the sentence frames, __10__ **plus** __5__ **equals** __15__, and __10__ **minus** __5__ **equals** __5__. Have partners write the number sentences.

Leveling: Emerging Read and model the question and answer. Children will repeat the answer.

Children at the Emerging level may have difficulty answering the question in a complete sentence, so allow them to respond with just the greater number as they gain confidence and skill.

partners 15 minutes

STRATEGY Model Language

Compare Two Shapes

Level: Expanding

Skills: Listening, Speaking, Reading, Writing

Objective: Compare shapes by adding –er to the adjective

Materials: paper, pencil, ruler

Activity Select 5 plane or solid shapes for partners to draw. Then have children look at each other's papers and take turns comparing their shapes by completing the sentence frames.

Children should both say and write their comparative sentences.

My ________ (shape) **is** ________ (adjective). **Your** ________ (same shape) **is** ________ (adjective with -er). **Adjective word bank:** tall long short small wide big narrow large	For example: **My rectangle is wide.** **Your rectangle is wider.**

Make sure children's comparative sentences are true based on the pictures they draw. Also, check the spelling of the comparative adjectives, specifically for "wider," "bigger," and "larger."

Extension Display all children's papers at the front of the room. Have children take turns comparing 2 shapes. For example, **Marc's cube is shorter than Jeannette's cube,** or **Jeannette's cube is short, but Marc's cube is shorter.**

Leveling: Emerging Draw two of each shape on the board, and write the sentence frames as well as the word bank. Point to a shape and read the first sentence, for example, **This square is small.** Then have the children point and reply chorally, **That square is smaller.**

partners 10 minutes

STRATEGY Model Language

How Much Is There?

Level: Expanding
Skills: Listening, Speaking
Objective: Ask and answer questions about how much
Materials: drawings that show amount, classroom objects

Activity Display classroom objects or drawings that show amount to have children practice describing how much there is. Make sure to choose objects that can be expressed as a collective amount such as paper, paint, or chalk. Liquids or food such as pizza are also good examples.

Write the sentence frames and word bank. Point to a drawing. Then have children ask and answer questions about how much there is.

Word bank: none, very little, not much, some, a lot of

How much ___milk___ **is there?**

There is ___a lot of___ ___milk___.

Extension Have children ask and answer the question for real objects. Have them discuss how asking "how much" is different from asking "how many."

Leveling: Emerging Read and model the question and answer. Children will repeat the answer. Show children some examples of what each expression means to help them connect meaning to the phrases.

Children at the Emerging level may have difficulty answering the question in a complete sentence, so allow them to respond with just the amount as they gain confidence and skill.

small group 15 minutes

STRATEGY Model Language

Which Is Your Favorite?

Level: Bridging

Skills: Listening, Speaking, Reading, Writing

Objective: Ask and answer questions to decide a favorite

Materials: drawings that show collections, classroom objects

Activity Display words, drawings, or objects that are a collection of 4–5 similar things such as fruits, vegetables, or toys. Have children practice describing which is their favorite.

Write the sentence frames. Point to a collection of objects and have children ask and answer questions about which is their favorite.

Which is your favorite __fruit__?

The __apple__ is my favorite.

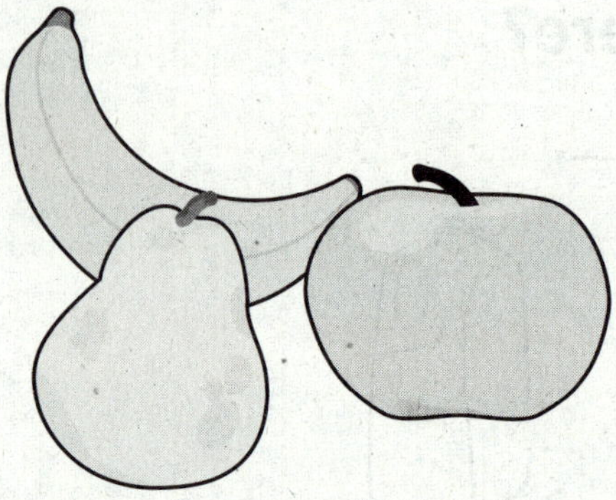

Have the other group members write the questions and answers as they are spoken.

Extension Have children survey the class, asking other children for their favorite from the collection. Children can make a tally table to show the results and then read the results, saying, **__Five__ children chose __pear__ as their favorite __fruit__.**

Leveling: Expanding Have children focus on reading and completing the sentence frames. Use a collection of 3 objects.

small group | 15 minutes

STRATEGY
Model Language

How Much Does Lunch Cost?

Level: Bridging
Skills: Listening, Speaking, Reading
Objective: Answer questions about how much lunch costs
Materials: food/price chart

Activity Give children a chart that lists the price of some lunch items.

Point to a picture or name a food item. Write the sentence frames and have children ask and answer questions about how much lunch costs.

What are you having for lunch?
How much does it cost?
I am having ___a hot dog___ **for lunch.**
Lunch costs ___$2.00___.

apple $0.25

hot dog $2.00

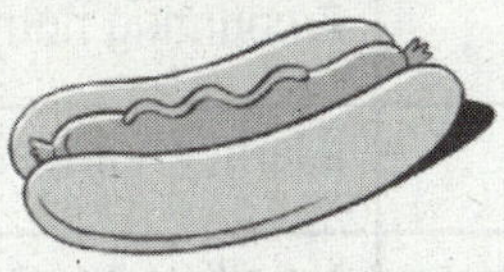

carrots $0.50

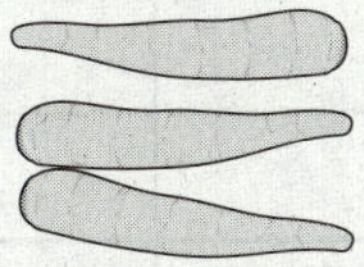

grapes $0.75

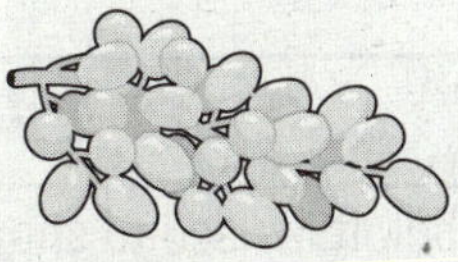

milk $0.60

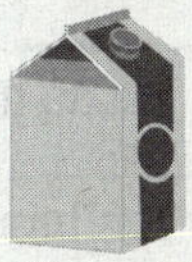

pizza $1.50

Have the other group members write the questions and answers as they are spoken.

Extension Children can ask about the combined price of objects. When children ask questions involving addition, they should ask and answer the question, **How much does** ___a slice of pizza___ **and** ___a milk___ **cost?** ___A slice of pizza___ **and** ___a milk___ **cost** ___$ 2.10___.

Leveling: Expanding Have children focus on reading and completing the sentence frames.

NUMBERS & OPERATIONS VOCABULARY

Use the EL definitions with children to help them understand math terms related to numbers and operations. The terms are organized by grade level so teachers can find vocabulary that applies to their children. For grade 1, children should also use grade K vocabulary. For grade 2, children should also use grade K and grade 1 vocabulary.

Grade	English	Spanish Cognate	Definition
K	add		To add is to join two groups and find a total.
K	are left		the remaining number or amount after subtraction
K	eight (8)		A counting number that is one more than seven.
K	eighteen (18)		A counting number that is ten and eight more.
K	eleven (11)		A counting number that is ten and one more.
K	equal sign (=)	signo de igualdad	The equal sign is a symbol. It is used to show "is equal to."
K	fewer		a smaller number or amount
K	fifteen (15)		A counting number that is ten and five more.
K	five (5)		A counting number that is one more than four.
K	four (4)		A counting number that is one more than three.
K	fourteen (14)		A counting number that is ten and four more.
K	greater		a larger number or amount
K	in all		"In all" is another way to say "total."
K	join		"Join" is another way to say "add."
K	less		not as much
K	minus (–)	menos	The minus sign is a symbol. It is used to show subtraction.
K	more		a greater quantity or amount
K	nine (9)		A counting number that is one more than eight.
K	nineteen (19)		A counting number that is ten and nine more.

Teacher's Tips

Children will probably be able to recite number names to 20 when counting, but may have difficulty with irregular number names such as eleven, twelve, and the teen numbers when actually working with them as tens and ones.

Many numbers and number concepts are best demonstrated with physical objects or visualized with drawings. Provide drawings and demonstrations whenever possible to help children gain understanding of the definitions.

Grade	English	Spanish Cognate	Definition
K	one		The first counting number.
K	one fewer		a number or amount that is one less
K	one more		a number or amount that is one greater
K	plus (+)		The plus sign is a symbol. It is used to show addition.
K	separate	separarse	To separate is to take away from a group.
K	seven (7)		A counting number that is one more than six.
K	seventeen (17)		A counting number that is ten and seven more.
K	six (6)		A counting number that is one more than five.
K	sixteen (16)		A counting number that is ten and six more.
K	subtract		To subtract is to take away objects from a group, or find out how many are left.
K	ten (10)		A counting number that is one more than nine.
K	thirteen (13)		A counting number that is ten and three more.
K	three (3)		A counting number that is one more than two.
K	twelve (12)		A counting number that is ten and two more.
K	twenty (20)		A counting number that is ten and ten more.
K	two (2)		A counting number that is one more than one.
K	zero (0)	cero	Zero is a number meaning none.
1	addend		Addends are any of the numbers that are added.
1	addition sentence		An addition sentence is a number sentence in which one number is added to another.
1	compare	comparar	You compare when you tell if numbers are equal to, less than, or greater than each other.
1	count back		To count back is to count backward from a given number.
1	count backward		This is a counting pattern in which each number is one less than, or just before, the previous number.

Teacher's Tips

Children may have difficulty distinguishing between the words *add* and *addend* and in recognizing which numbers in an addition problem are the addends and which number is the sum.

Grade	English	Spanish Cognate	Definition
1	count forward		This is a counting pattern in which each number is one more than, or just after, the previous number.
1	count on		To count on is to count forward from a given number.
1	difference	diferencia	The difference is the answer to a subtraction problem.
1	doubles	dobles	This is an addition fact. The two addends are the same.
1	doubles minus one		This is a near doubles fact. One addend is one less than the other; $6 + 5 = 11$.
1	doubles plus one		This is a near doubles fact. One addend is one greater than the other; $6 + 7 = 13$.
1	even		An even number of objects can be grouped in pairs, or twos.
1	fact family		A fact family uses the same numbers to make addition and subtraction sentences.
1	greatest		the largest in size; the most in amount
1	hundred (100)		A counting number. It is equal to 10 tens or 100 ones.
1	is equal to (=)	es igual a	has the same value
1	is greater than (>)		greater than; larger in size, or more in amount
1	is less than (<)		less than; smaller in size, or fewer in amount
1	least		fewest in amount
1	make a ten		This is a way to break down the addends to make a ten plus ones fact; $9 + 7 = 9 + 1 + 6 = 10 + 6$
1	number line	línea numerada	A number line is a line marked with a series of numbers. The numbers are evenly spaced along the line.
1	one less		a number or amount that is one fewer

Teacher's Tips

The use of the terms *tens* and *ones* as place value can be confusing. Children may also confuse *one* and *won* when these words are spoken.

Children may be able to make numbers and recognize hundreds, tens, and ones digits, but have trouble articulating what they have done. Conversely, they may be able to say the vocabulary, but be unable to apply it properly. Review each term. Make sure children understand that a hundred consists of ten tens, and a ten consists of ten ones. The number 10 is one ten. Use tens blocks to illustrate as you count by tens.

Grade	English	Spanish Cognate	Definition
1	ones		The ones place on a place-value chart tells how many ones are in a number.
1	order	orden	Order is the way in which things are placed.
1	part	parte	A part is one part of a whole. In 3 + 1 = 4, 3 and 1 are both parts of the whole.
1	related facts		Related facts are addition and subtraction facts that use the same numbers.
1	skip count		To skip count is to count in a pattern that skips numbers.
1	subtraction sentence		A subtraction sentence is a number sentence in which one number is subtracted from another.
1	sum	suma	The sum is the answer to an addition problem.
1	take away		To take away is to remove or subtract from a group.
1	tens		The tens place on a place-value chart tells how many tens are in a number.
1	whole		all of a single item or group
2	cent sign (¢)	signo de centavo	The cent sign is a symbol. It represents cents.
2	decimal point	punto decimal	A decimal point is a period used to separate dollars from cents.
2	digits	dígitos	Digits are the symbols 0, 1, 2, 3, 4, 5, 6, 7, 8, and 9. They can have different place values, depending where they are in a number.
2	dime		A dime is a coin worth 10 cents (10¢ or $0.10). A dime is equal to 10 pennies.

Teacher's Tips

Children may have difficulty with *sum* and *some*, which sound the same but have different meanings and spellings.

In some Spanish-speaking countries, the symbols separating digits to indicate place value (the comma and the decimal point) are used in a different way than they are in the United States. For example, the number 1,906.43 in the United States is written as 1.906,43 in some Spanish-speaking countries.

The difference between a number and a digit may be difficult for children to grasp. Give children examples. For example, in the number 48, 4 and 8 are the digits, just as they are in the number 84. The digits are 0, 1, 2, 3, 4, 5, 6, 7, 8, 9. All numbers are made up of these digits, just as all English words are made up of the 26 letters of the English alphabet. A digit's value is dependent on its place.

Grade	English	Spanish Cognate	Definition
2	dollar	dólar	A dollar is a coin or bill worth 100 cents ($1.00).
2	dollar sign ($)	signo de dólar	The dollar sign is a symbol. It stands for a dollar.
2	equal parts	partes iguales	Equal parts are parts of an object or group that have been divided equally into pieces.
2	estimate	estimar, estimación	To estimate is to find about how many or about how much.
2	even number		An even number is a whole number that can be divided by 2 or shown as pairs.
2	fourths		four equal parts
2	half dollar		A half dollar is a coin worth 50 cents (50¢ or $0.50). A half dollar is equal to 50 pennies.
2	halves		two equal parts
2	hundreds		The hundreds place on a place-value chart tells how many hundreds are in a number.
2	nickel		A nickel is a coin worth 5 cents (5¢ or $0.05). A nickel is equal to 5 pennies.
2	odd		An odd number is a whole number that cannot be divided by 2.
2	penny		A penny is a coin worth 1 cent (1¢, or $0.01).
2	quarter		A quarter is a coin worth 25 cents (25¢ or $0.25). A quarter is equal to 25 pennies.
2	regroup	reagrupar	When you regroup, you change a number from one form into an equal form. For example, when subtracting, you might exchange one ten for ten ones.
2	rule		A rule describes how to make an 'output.' Rules use addition and subtraction.
2	thirds		three equal parts
2	thousand (1;000)		There are 10 hundreds in 1 thousand.

MEASUREMENT VOCABULARY

Use the EL definitions with children to help them understand math terms related to measurement. The terms are organized by grade level so teachers can find vocabulary that applies to their children. For grade 1, children should also use grade K vocabulary. For grade 2, children should also use grade K and grade 1 vocabulary.

Grade	English	Spanish Cognate	Definition
K	afternoon		Afternoon is the part of the day from noon until sunset.
K	day		A day is a unit of time equal to 24 hours.
K	heavier		A heavier object weighs more than another object.
K	heaviest		The heaviest object weighs the most.
K	less time		Something that takes less time does not take as long.
K	lighter		A lighter object is easier to lift.
K	lightest		The lightest object weighs the least.
K	longer		Something that is longer has more length than another object
K	longest		The longest object has the most length.
K	month		A month is one of the twelve time sections a year is divided into.
K	more time		Something that takes more time takes longer.
K	morning		Morning is the part of the day from sunrise to noon.
K	shorter		Something that is shorter is not as long as something else.
K	shortest		The shortest object has the least length.
K	taller		Something that is taller has more height than another object.
K	tallest		The tallest object or person has the most height.
K	today		Today is the present day.

Teacher's Tips

Many measurement concepts are best demonstrated with physical objects or visualized with drawings. Provide drawings and demonstrations whenever possible to help children gain understanding of the definitions.

Grade	English	Spanish Cognate	Definition
K	tomorrow		Tomorrow is the day after today.
K	week		A week is a unit of time that equals seven days.
K	year		There are 12 months in a year.
K	yesterday		Yesterday is the day before today.
1	balance		A balance is a tool used to compare weight.
1	capacity	capacidad	Capacity is the amount a container can hold.
2	centimeter (cm)	centímetro	A centimeter is a metric unit of measurement. It is used to measure length or distance.
2	cup		A cup is a customary unit of measurement. It is used to measure capacity.
2	foot (ft)		A foot is a customary unit of measurement. It is used to measure length or distance. 1 foot equals 12 inches.
2	gallon (gal)	galón	A gallon is a customary unit of measurement. It is used to measure capacity. 1 gallon equals 4 quarts.
2	gram (g)	gramo	A gram is a metric unit of measurement. It is used to measure mass.
2	half hour		A half hour is 30 minutes.
2	hour	hora	An hour is a unit used to measure time. 1 hour equals 60 minutes.
2	hour hand		The hour hand is the the short hand on an analog clock.
2	inch (in.)		An inch is a customary unit of measurement. It is used to measure length or distance.
2	kilogram (kg)	kilogramo	A kilogram is a metric unit of measurement. It is used to measure mass. 1 kilogram equals 1,000 grams.
2	liter (L)	litro	A liter is a metric unit of measurement. It is used to measure capacity. 1 liter equals 1,000 milliliters.

Teacher's Tips

Children are likely to be familiar with the word *cups*, meaning "containers people drink out of." Explain that there are different sizes of drinking cups, and that drinking cups can hold different amounts of liquid. Then explain that a *cup* as a unit of measure is always a specific size.

Children may notice that the abbreviation for *liter* is shown as a capital letter, while the other abbreviations are lowercase. Explain to them that this is done to distinguish it from the number 1, which a lowercase l resembles.

Grade	English	Spanish Cognate	Definition
2	mass	masa	Mass is the amount of matter an object has.
2	meter (m)	metro	A meter is a metric unit of measurement. It is used to measure length or distance. 1 meter equals 100 centimeters.
2	milliliter (mL)	mililitro	A milliliter is a metric unit of measurement. It is used to measure capacity. 1,000 milliliters equal 1 liter.
2	minute (min)	minuto	A minute is a unit used to measure short amounts of time. There are 60 minutes in 1 hour.
2	minute hand		The minute hand is the long hand on an analog clock.
2	ounce (oz)	onza	An ounce is a customary unit of measurement. It is used to measure weight. 16 ounces equal 1 pound.
2	pint (pt)	pinta	A pint is a customary unit of measurement. It is used to measure capacity. 1 pint equals 2 cups.
2	pound (lb)		A pound is a customary unit of measurement. It is used to measure weight. 1 pound equals 16 ounces.
2	quart (qt)	cuarto	A quart is a customary unit of measurement. It is used to measure capacity. 1 quart equals 2 pints.
2	volume	volumen	Volume is the amount of space a solid figure takes up.
2	yard (yd)	yarda	A yard is a customary unit of measurement. It is used to measure length or distance. 1 yard equals 3 feet.

Teacher's Tips

Children should recognize how most of the measurement abbreviations correspond to their spelled-out names. However, children may be confused by the abbreviation for pound, *lb*, as it bears no resemblance to the word it represents. You might explain to children that the abbreviation *lb* comes from the Latin word for *pound,* which is *libra.*

GEOMETRY VOCABULARY

Use the EL definitions with children to help them understand math terms related to geometry. The terms are organized by grade level so teachers can find vocabulary that applies to their children. For grade 1, children should also use grade K vocabulary. For grade 2, children should also use grade K and grade 1 vocabulary.

Grade	English	Spanish Cognate	Definition
K	circle	círculo	A circle has a round shape.
K	corner		A corner is the point where sides meet.
K	cube	cubo	A cube is a solid shape with six square faces.
K	curve	curva	A curve is rounded.
K	curved surface		A curved surface is a rounded surface.
K	cylinder	cilindro	A cylinder is a solid shape. It has two bases that look like circles. It also has a curved surface.
K	flat surface		A flat surface is the outside of an object. It can be the top, bottom, or face of an object.
K	hexagon	hexágono	A hexagon is a solid shape with six sides.
K	pattern	patrón	A pattern is a series that follows a rule that repeats.
K	rectangle	rectángulo	A rectangle has four sides. It has two pairs of equal sides. It has four corners.
K	rhombus	rombo	A rhombus has four sides. All of the sides are the same length.
K	roll		To roll is to move an object like a ball along its curved surface.
K	shape		The shape is the outline of an object.
K	side		A side is one of the line segments that forms a shape.
K	slide		To slide is to move an object across a surface without picking it up.
K	sort		To sort is to group objects that are alike in some way.
K	sphere	esfera	A sphere is a round solid shape such as a ball.
K	square		A square has four equal sides. It has four corners.
K	stack		To stack is to put one object on top of another.

Teacher's Tips

Many shapes and geomtry concepts are best demonstrated with physical objects or visualized with drawings. Provide drawings and demonstrations whenever possible to help children gain understanding of the definitions.

Grade	English	Spanish Cognate	Definition
K	trapezoid	trapezoide	A trapezoid has four sides. It has one pair of parallel sides.
K	triangle	triángulo	A triangle has three sides. It has three corners.
K	vertex/vertices	vértice/vértices	A vertex is a point where sides meet.
1	cone	cono	A cone is a solid shape. It has a base that looks like a circle. It comes to a point at the top.
1	edge		An edge is the line segment where two faces of a solid shape meet.
1	face		A face is a flat surface of a three-dimensional shape.
1	growing pattern		A growing pattern is a pattern in which the number or figure gets larger by the same amount each time.
1	pyramid	pirámide	A pyramid has a base that looks like a square. It has four faces that look like triangles. The faces come to a point at the top.
1	rectangular prism	prisma rectangular	A rectangular prism is a box. Each flat surface is a rectangle.
1	repeating pattern	patrón que se repite	A repeating pattern is a pattern in which the pattern unit is used over and over again.
1	pattern unit	unidad de patrón	A pattern unit is a sequence of elements that are repeated over and over.

Teacher's Tips

Terms like *cone/cono, vertex/vértice,* and *pyramid/pirámide* are cognates. If students are familiar with the Spanish term, help them relate it to the English term by pointing out similarities and differences between the Spanish word and the English word.

The word *face* has multiple meanings. Find out what students think the term means. Where possible, relate the precise math meaning to one that students already know.

SPATIAL RELATIONSHIPS VOCABULARY

Use the EL definitions with children to help them understand math terms related to spatial relationships. The terms are organized by grade level so teachers can find vocabulary that applies to their children. For grade 1, children should also use grade K vocabulary. For grade 2, children should also use grade K and grade 1 vocabulary.

Grade	English	Spanish Cognate	Definition
K	above		Something that is above is overhead or in a higher place.
K	after		If something is after, it comes behind in place or order.
K	before		If something is before, it is ahead in place or order.
K	below		Something that is below is beneath or in a lower place.
K	beside		at the side of; next to
K	between		Something that is between is in the middle. 12 is between 11 and 13.
K	inside		the inner part of something
K	left		a direction; the opposite of right
K	next to		beside
K	outside		the outer part of something
K	over		Something that is over is overhead or in a higher place.
K	right		a direction; the opposite of left
K	under		Something that is under is below.

Teacher's Tips

Many spatial relationships are best demonstrated with physical objects or visualized with drawings. Provide drawings and demonstrations whenever possible to help children gain understanding of the definitions.

Name ______________________________

Draw the hands of the clock.
Write the time below each clock.

:

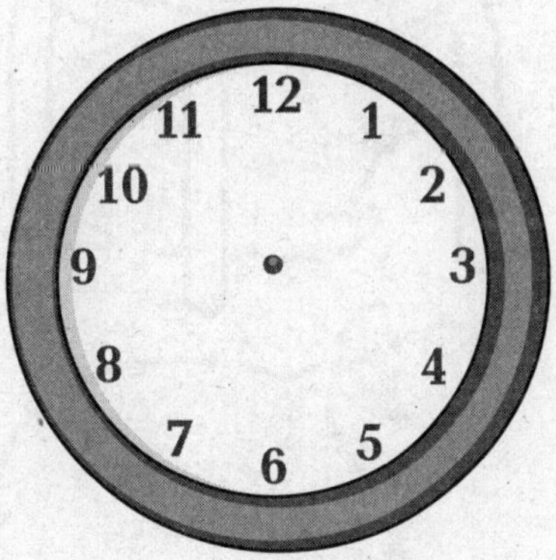

:

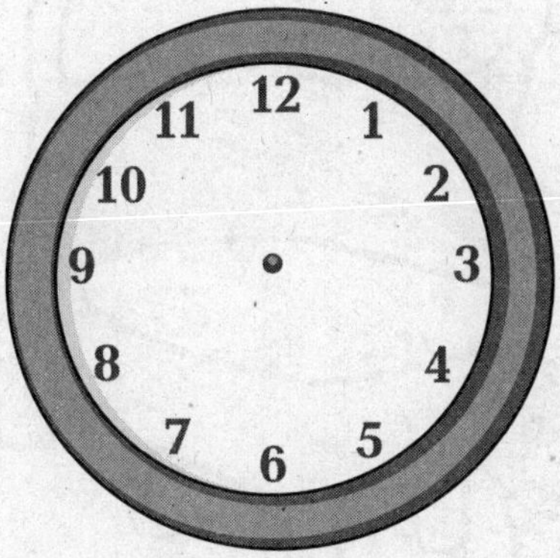

:

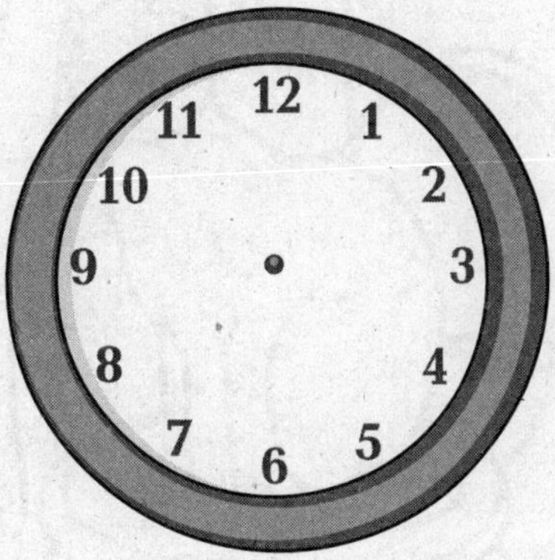

:

Name ______________________________

Name ______________________________

Name ______________________________

We can find the difference. We can get the sum.
We can use subtraction and addition.

When we add 2 and 2 and get 4, we're done,
'cause we get the answer. We get the sum.

When we subtract 4 from number 11,
We find the difference; it's number 7.

We use the plus sign to add 5 and 2.
We use the plus sign 'cause we have to.

When we subtract 5 from number 9,
We use the minus sign every time.

We use minus to subtract and plus to add.
Finding the sum or difference isn't so bad.

Name ____________________

I can use doubles plus one
to get the sum, to get the sum.
I can use doubles plus one
to get the sum in addition.

How can I use doubles plus one
to get the sum, to get the sum?

I add the doubles and then add one
to get the sum, to get the sum.
I add the doubles and then add one
to get the sum, to get the sum.

Name ______________________________

Can you change the order of addends?
Will the sum be the same?
Yes, you can change the order of addends
and the sum will be the same.

You want to find the sum
of the addends 3 and 1.
Is the answer 4?
Yes, the sum is 4.

If you want to see
the sum of 1 and 3,
is the answer still 4?
Yes, the sum is still 4.

Can you change the order of addends?
Will the sum be the same?
Yes, you can change the order of addends
and the sum will be the same.

Name ______________________________

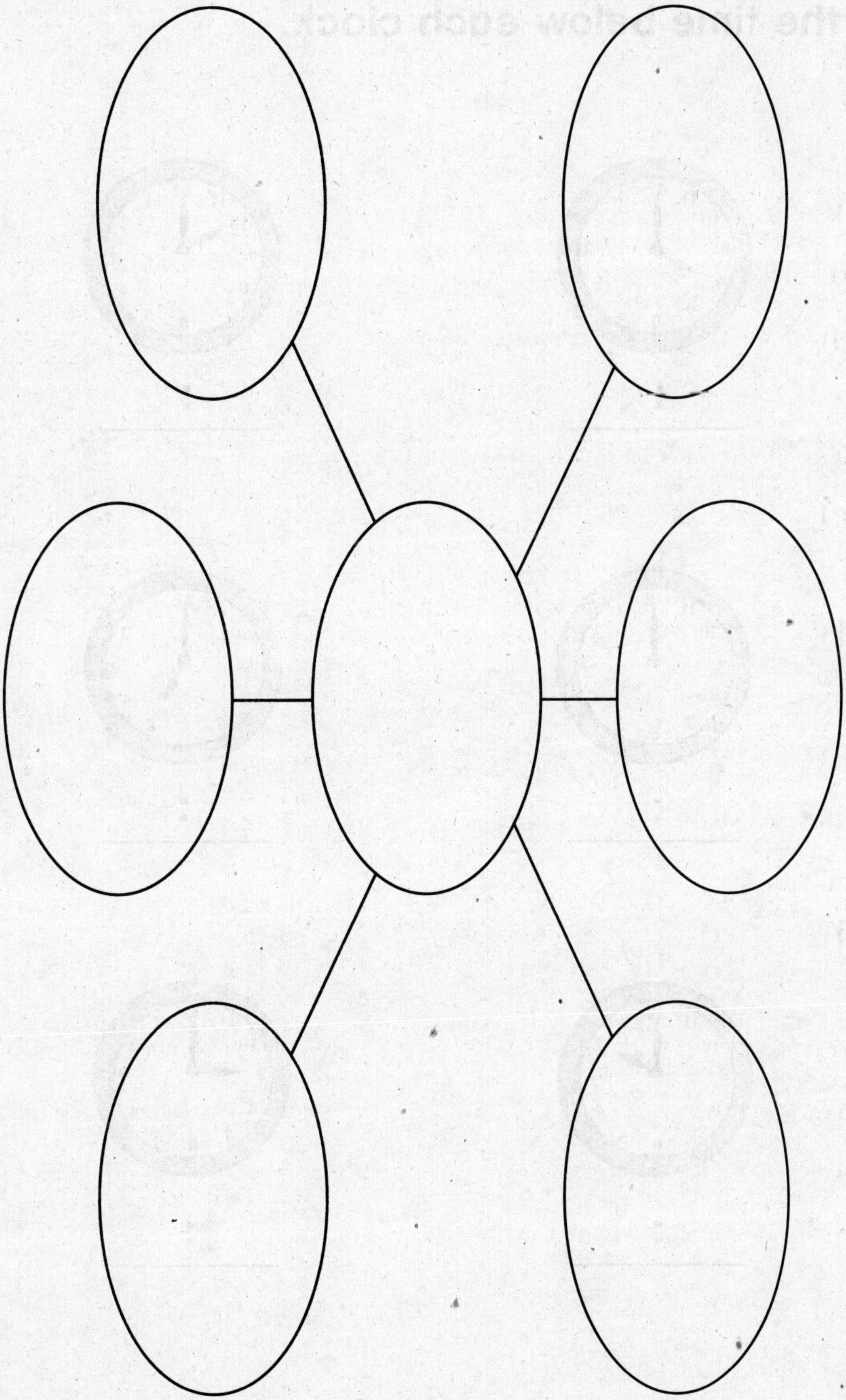

Name ____________________

Write the time below each clock.

: ________

: ________

: ________

: ________

: ________

: ________

Name ______________________________

Write the time below each clock.

:

:

:

:

:

:

Name ______________________________

My brother John and I
are about the same size,
and we both have
big brown eyes.

We're about the same height
and I think that's fine,
because he is four foot eight
and I am four foot nine.

We're about the same weight,
as you can see:
he weighs 52 pounds
and I weigh 53.

Yes, my brother John and I
are about the same size,
and he's the best brother
in my eyes.

Name ______________________________

**Match the words on the left to the pictures on the right.
Write the word next to the picture.**

1. dog ●

2. cat ●

3. horse ●

4. bird ●

5. fish ●

6. cow ●

7. rabbit ●

8. mouse ●

●  ______________

●  ______________

●  ______________

● 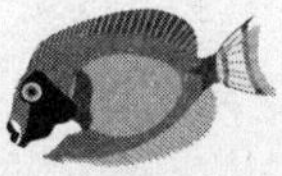 ______________

●  ______________

●  ______________

●  ______________

●  ______________

Name ______________________

Match the words on the left to the pictures on the right.
Write the word next to the picture.

1. pants ● ●  ______

2. shoes ● ●  ______

3. shirt ● ●  ______

4. cap ● ●  ______

5. dress ● ●  ______

6. skirt ● ●  ______

7. socks ● ●  ______

8. jacket ● ● 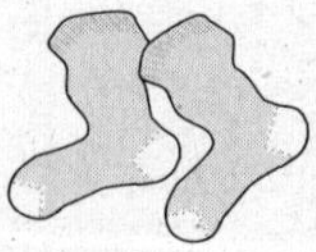 ______

Name ______________________________

Match the words on the left to the pictures on the right.
Write the word next to the picture.

1. banana ● ●  ____________
2. apple ● ● 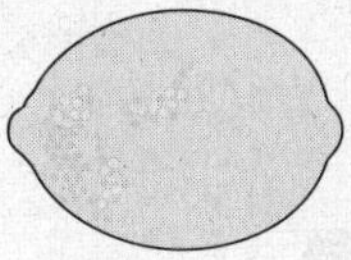 ____________
3. orange ● ● 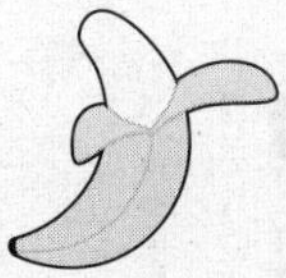 ____________
4. lemon ● ● 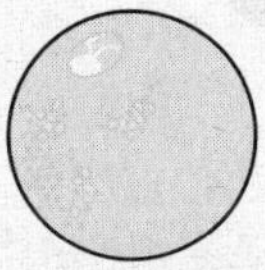 ____________
5. pineapple ● ● 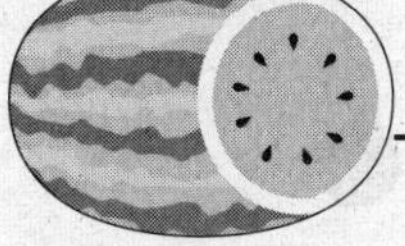 ____________
6. watermelon ● ●  ____________

Name ______________________________

Match the words on the left to the pictures on the right.
Write the word next to the picture.

1. pumpkin ● ●  ______________

2. pepper ● ●  ______________

3. tomato ● ●  ______________

4. radish ● ●  ______________

5. pea pod ● ●  ______________

6. carrot ● ● 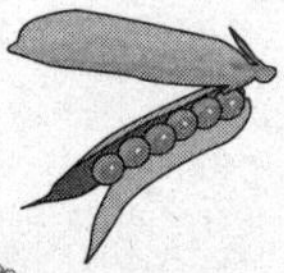 ______________

7. bean ● ● 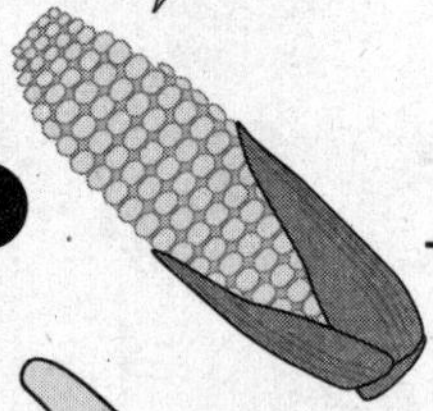 ______________

8. corn ● ● 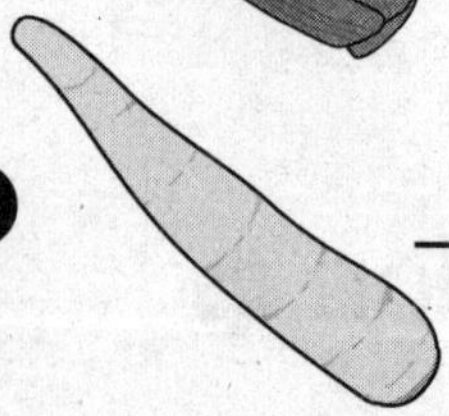 ______________

Name ____________________

Match the words on the left to the pictures on the right.
Write the word next to the picture.

zero ●	● 1 ________
one ●	● 3 ________
two ●	● 2 ________
three ●	● 0 ________
four ●	● 5 ________
five ●	● 4 ________
six ●	● 8 ________
seven ●	● 6 ________
eight ●	● 10 ________
nine ●	● 7 ________
ten ●	● 9 ________

Name ______________________________

Match the words on the left to the pictures on the right.
Write the word next to the picture.

1. cone ●

2. cylinder ●

3. sphere ●

4. pyramid ●

5. cube ●

●  ____________

●  ____________

● 

____

●  ____________

●  ____________

Name ______________________________

Match the words on the left to the pictures on the right.
Write the word next to the picture.

1. triangle ● ●  ______________

2. square ● ●  ______________

3. rectangle ● ●  ______________

4. rhombus ● ● 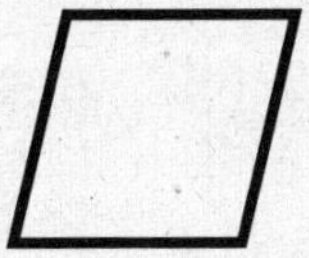 ______________

5. circle ● ● 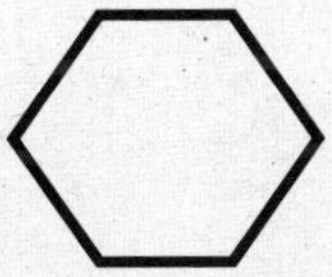 ______________

6. trapezoid ● ●  ______________

7. hexagon ● ● 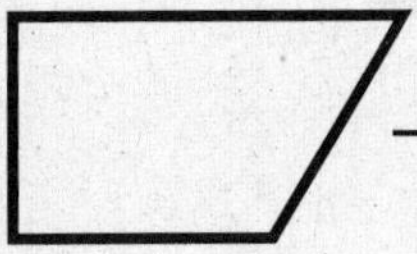 ______________

Name ____________________________

**Match the words on the left to the pictures on the right.
Write the word next to the picture.**

1. penny ● ● 

____

2. nickel ● ● 

____

3. dime ● ● 

____

4. quarter ● ● 

____

5. dollar ● ● 

____

Name ______________________________

Read the poem.
Draw the shapes.

I can name shapes, and colors too.
That is a square and it is blue.

That is a triangle and I have seen
it is in the color green.

For this one I have to scratch my head
but then I know it is a rectangle and it is red!

That circle looks just like the sun
and that circle is an orange one.

That one looks like a square, I think.
No, it is a rhombus and it is pink!

I know the names of shapes, and colors too.
Are there any others I can name for you?

Time to the Hour

Name ______________________

Write the time below each clock.

8:00

10:00

12:00

5:00

2:00

9:00

Time to the Half Hour

Name ______________________

Write the time below each clock.

1:30

5:30

8:30

10:30

3:30

12:30

Matching: Animals

Name ______________________

Match the words on the left to the pictures on the right. Write the word next to the picture.

1. dog
2. cat
3. horse
4. bird
5. fish
6. cow
7. rabbit
8. mouse

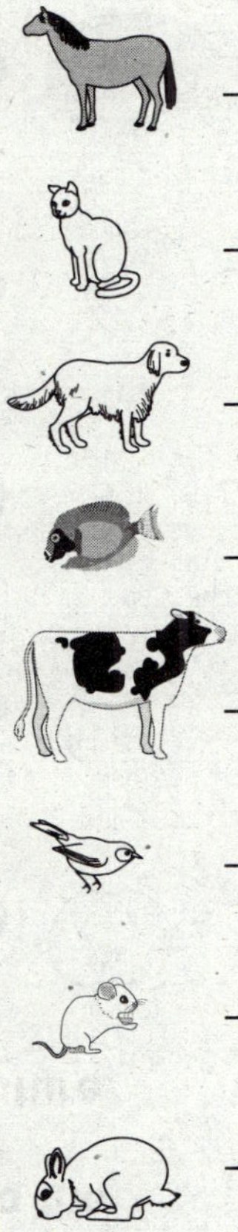

	horse
	cat
	dog
	fish
	cow
	bird
	mouse
	rabbit

Matching: Clothing

Name ______________________

Match the words on the left to the pictures on the right. Write the word next to the picture.

1. pants
2. shoes
3. shirt
4. cap
5. dress
6. skirt
7. socks
8. jacket

	pants
	shirt
	cap
	shoes
	skirt
	jacket
	dress
	socks

Name ______________________

Match the words on the left to the pictures on the right.
Write the word next to the picture.

1. banana
2. apple
3. orange
4. lemon
5. pineapple
6. watermelon

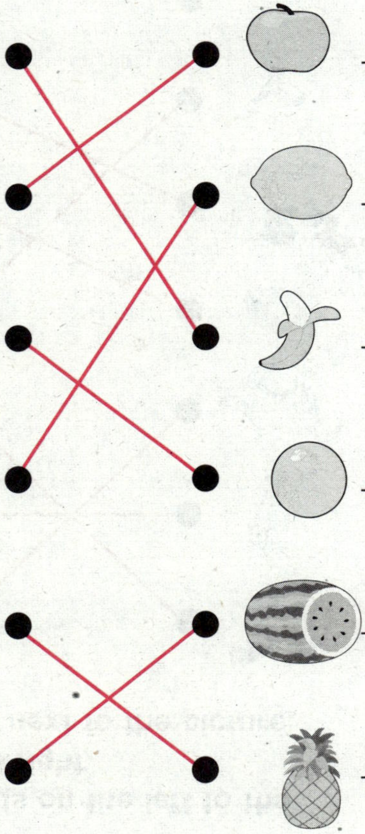

apple

lemon

banana

orange

watermelon

pineapple

Name ______________________

Match the words on the left to the pictures on the right.
Write the word next to the picture.

1. pumpkin
2. pepper
3. tomato
4. radish
5. pea pod
6. carrot
7. bean
8. corn

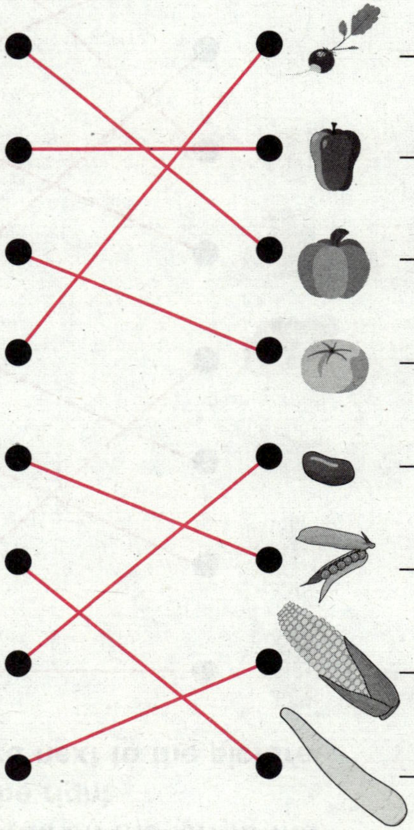

radish

pepper

pumpkin

tomato

bean

pea pod

corn

carrot

Matching: Numbers

Name ______________________

Match the words on the left to the pictures on the right.
Write the word next to the picture.

Words	Pictures	Answers
zero	1	one
one	3	three
two	2	two
three	0	zero
four	5	five
five	4	four
six	8	eight
seven	6	six
eight	10	ten
nine	7	seven
ten	9	nine

Matching: Solid Shapes

Name ______________________

Match the words on the left to the pictures on the right.
Write the word next to the picture.

1. cone		sphere
2. cylinder	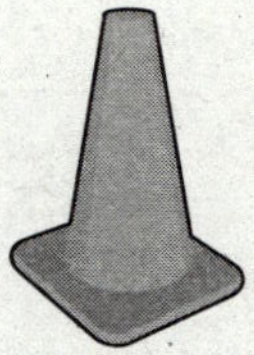	cone
3. sphere		cylinder
4. pyramid		cube
5. cube		pyramid

Matching: Plane Shapes

Name ______________________

Match the words on the left to the pictures on the right.
Write the word next to the picture.

1. triangle
2. square
3. rectangle
4. rhombus
5. circle
6. trapezoid
7. hexagon

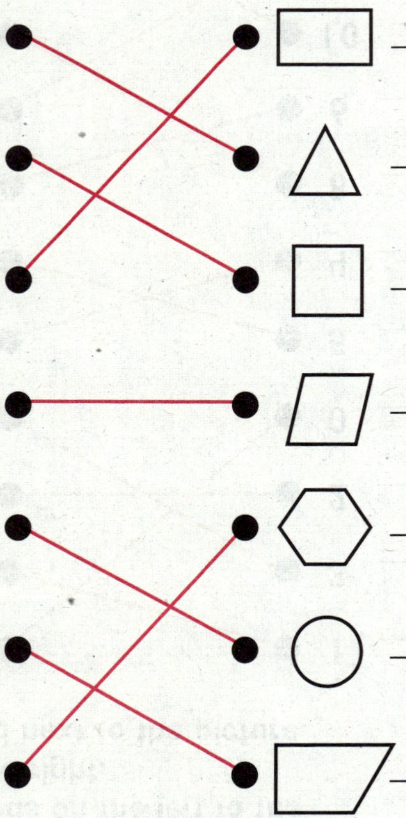

rectangle
triangle
square
rhombus
hexagon
circle
trapezoid

Matching: Money

Name ______________________

Match the words on the left to the pictures on the right.
Write the word next to the picture.

1. penny
2. nickel
3. dime
4. quarter
5. dollar

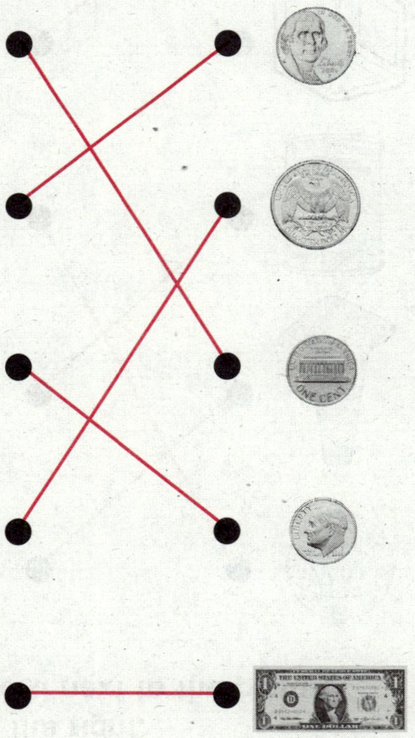

nickel
quarter
penny
dime
dollar